Technology and Social Choices in the Era of Social Transformations

Matej Makarovič / Borut Rončević (eds.)

# Technology and Social Choices in the Era of Social Transformations

**PETER LANG**

**Bibliographic Information published by the
Deutsche Nationalbibliothek**
The Deutsche Nationalbibliothek lists this publication in the Deutsche
Nationalbibliografie; detailed bibliographic data is available online at
http://dnb.d-nb.de.

**Library of Congress Cataloging-in-Publication Data**
A CIP catalog record for this book has been applied for at the
Library of Congress.

This book was co-funded and supported by the National Committee of the Management of Social Transformations Programme at the Slovenian National Commission
for UNESCO and by the Slovenian Research Agency, grant number J5—1788.

ISBN 978-3-631-80821-4 (Print)
E-ISBN 978-3-631-83700-9 (E-PDF)
E-ISBN 978-3-631-83701-6 (EPUB)
E-ISBN 978-3-631-83702-3 (MOBI)
DOI 10.3726/b17652

# Contents

6                              Contents

# Inseparability of Technology and Society: An Introduction

Technology cannot be separated from society. From the dawn of humanity, the dialectic relationship between the two has been one of the driving forces behind changes in both realms. Trends in technological developments and their applications are, ultimately, the result of individual and collective choices. At the same time, technology influences the social choices of individuals, small groups and entire societies.

As a result, it is impossible to avoid the issue of technology in an analysis of the development of societies. These developments are, whether examining them at a grand scale or in long-term perspectives, often analysed as the global trends or megatrends. There have been many attempts at mapping them. One such approach, from the perspective of sociology, is the analysis of the four global trends by Genov (1997). None of the four trends would be possible without technological developments. The trend of the spread of instrumental activism is best demonstrated through the ideology and practice of industrialism that have been incorporated in technological infrastructures. The trend of individualization is immeasurably enabled by technological devices and enhanced by the rapid pace of technological change, which makes certainties that were previously granted by the levels of collective-based educational processes unattainable. The trend of upgrading organizational rationality is increasingly inseparable from technological innovations, pioneered in their industrial settings in Western Europe, and then spread throughout the globe. Even the trend of value-normative universalization would not be possible without technological advances, be it the Guttenberg's invention of the printing press or the internet. Since communication is mediated through technology – be it papyrus scrolls or online social media – it goes without saying that the challenges to these trends (Genov, 2018) also cannot be separated from technological development. Technological development, together with 'crises of capitalism', may, in fact, amplify them.

Likewise, a review of Naisbitt's ten megatrends (1982) clearly shows that these trends were formed in a dialectic relationship between society and technology, which is not only true for the most obvious trends, such as the transition from an industrial to an information society or the shift from forced technology to high-tech. Technology also enables other trends or, in some cases, trends require the development of specific technological solutions. The emergence of a truly global economy is only possible through the emergence of extremely sophisticated technology-enabled (both IT and transportation technology) logistical value chains, by extremely sophisticated infrastructure such as data centres and, in some cases, even by the joint development of products, again enabled by the infrastructure. The shift in business management from short-term planning to long-term perspectives relies on increasingly sophisticated tools, all of which are enhanced by technology. Even the less obvious ones, such as the shift from institutional help, provided by the government, medical institutions, the school system, and similar, to self-help, are enabled by – and require the development of – relevant technological tools, such as the ambient assisted living devices, smart homes, online medical assistance, a plethora of online training services, and similar.

The unavoidability of this dialectic relationship is also exemplified in more immediate concerns of strategists and policy-makers. While one of the key goals of the key strategic document of the European Union was growth, it was recognized that this growth needs to be smart, inclusive and sustainable, all of which require significant investments in technological development. This goal was then operationalized through the Smart Specialisation Strategy approach (Asheim et al., 2015, p. 23; Foray, 2014) and is currently a key factor in the bottom-up process commonly recognized as Industry 4.0 (Lasi et al., 2014).

The topic is, therefore, ubiquitous and as relevant as ever. This book is a collection of papers addressing the central topic of technology and social choices from a variety of perspectives. It thus provides a series of inter-connected insights on how the rapid developments of digitalization and related technologies affect the choices people are making at the micro- (individuals and groups), mezzo- (academic institutions, companies and associations) and macro- (national level governance) levels, and in different social spheres or subsystems: media, education, economy and politics.

While we include contributions by different authors using a variety of theoretical perspectives and methodological approaches, the book connects them by unfolding a story of technologically mediated communication taking place through the media and the formal educational system. We relate this communication to the potentials for the activation of people, which may lead to establishing human agency contributing to civic participation and innovative potentials, observable in the community and entrepreneurial practices. However, this may lead to disorientation and passivity manifested through some macro-level social choices, such as the rise of populism or the persistence of authoritarianism.

The contributions in this volume are interconnected through two closely related focuses: technological development and social choices. While relating them, we assume the relationship between the three components: (1) human individuals and their agency; (2) social structures, both as the initial context and as resulting from human agency; and (3) technology that has been developed and applied by human agents' choices within social contexts, while it also affects these contexts and future social choices. While each of the components has its own specific and emergent properties, they also trigger the processes of change in each other.

Each of the components is seen as contributing to the increased dynamics of the other two. The combination of social choices done by human agents and technological development has made social structures subject to change that takes place with a pace unprecedented by anything before in human history. We are moving from a morphostatic to an increasingly morphogenetic society (Archer, 2012; 2017).

This provides a distinct challenge for the individuals, which is mainly addressed in the first section of this book. Rapid social change combined with the increasingly technologically mediated environment prevents individuals from merely following the routines and practices tested by previous generations. Both the technological and the social environment are changing so quickly, thus contributing to new dimensions of generational gaps and further eroding the established inter-generational relationships. As shown by Mateja Rek in her contribution to this volume – comparing the media habits between youth people and senior citizens – the generational gap seems to be quite consistent with the digital gap. However, young people also face significant shortcomings in their ability to reflect on media contents that they are exposed to and to participate in the content creation process in a beneficial and desirable way (Rek, this volume).

Being increasingly less able to rely on the experience of the previous generations in an increasingly morphogenetic social and technological context, young people need to develop higher levels of reflexivity. Experiencing the global capitalist social order may have taught them to strive for individual performance in a highly competitive context. In contrast, the increasing need for sustainability and the maintenance of social cohesion also requires collaboration beyond the strictly economic and individualist logic – in terms of relational reflexivity. The research by Tea Golob and Matej Makarovič notes that different uses of digital technology can be linked, on the one hand, to the development of relational reflexivity enabling individuals to participate in the creation of the common (relational) goods but also, on the other, to personal disorientation.

While specific uses of digital technology can be empowering in terms of making people more (relationally) reflexive, they can also help in addressing psychosocial problems. The case presented by Jana Krivec, Tjaša Stepišnik and Primož Rakovec offers a practical example of a personalized online counsellor that combines virtual community, e-therapies and avatars. Clearly, digital technology can be both a part of the problem and a part of the solution.

It is often also seen as a solution for upgrading the capacities of the educational system. This is the topic of the chapter of Andrej Raspor and Petra Kleindienst, which opens the second section of the book – focused on learning and innovation. Here, the authors demonstrate how the learning process at the tertiary level, which increasingly takes place online, contributes to students' presentation skills.

In addition, the pace of social and technological change also requires going beyond the classical divisions between the time of study and time of work as well as between academia and entrepreneurship. The students studying on-line, presented in the above-mentioned research, are thus typically part-time students combining their study with work. In contrast, also in the production of knowledge, we are witnessing an increasing cognitive mobilization or the cognitization of society, as manifested, for example, through high tech and academic entrepreneurship – as presented in the contribution by Frane Adam, Maruša Gorišek and Matej Makarovič. They also refer to the issue of reflexivity as the 'society that is built and dependent on the principles of innovation also needs the ability to take a reflexive attitude to its future and the long-term coordination and synergy of social actors' (Adam et al., this volume).

Finally, macro-level coordination is addressed in the third and final section of the book, addressing political challenges. While the rise of digital media has strongly affected political life, there are also several other actors' choices and structural conditions affecting the political processes. Thus far, the cases from North Africa and the Middle East, presented by Janja Mikulan Kildi, have demonstrated the persistence of authoritarianism as well as the continuing relevance of the military leaders as crucial social actors in these contexts. Global interdependencies and technological development have not provided any kind of apparent global political convergence in terms of democratization. 'The End of History' (Fukuyama, 1989), in terms of globalized liberal democracy, has turned out to be an illusion.

Even more so, Adam and Tomšič stress the crisis of existing democracies as one of the causes for the rise of political populism and the personalization of politics in European and other democracies. However, the authors also note the diversity and complexity of the populist phenomena. Finally, Dadiana Chiran confronts the economic and cultural factors of voters' choices and reviews a series of models supposed to predict voting behaviour not only in the old but also in the new democracies.

It may be added that the rise of populism may also be related to reflexivity failures – where actors are enabled to reach and/or implement post-reflexive choices to respond to the rising globalized societal and technological challenges. Thus, they may resort to seemingly attractive yet clearly oversimplified 'solutions'.

The context of globalization is at least implicitly present in all of the contributions in the volume. Even when presented through specific national case studies, as in most of the chapters in this book, all of the findings should be seen in the global context. Active life in modern society requires both living with rapid technological development and with the globalized societal context. It is thus not surprising that instead of following the footsteps of their ancestors, young people of today are expected to be technologically proficient and internationally mobile to be full members of the emerging social order (Golob, 2017). However, these expectations also create divisions and exclusion of those left behind in the globalized and digitalized contexts – thus again offering fertile ground for the mobilization in favour of the populist political actors.

Finally, the contributions to this volume were all written before the Covid-19 pandemic that will definitely affect the local, national and global social contexts for the foreseeable future. However, the new situation has not affected the relevance of the major points of the contributions in this book. The significance of digital technology for experiencing the world, education and innovation might become even greater in this context. However, to counter simplified populist or authoritarian paths, higher levels of reflexivity will also be needed to address the new situation, especially in terms of establishing more sustainable developmental models – in terms of balancing individual freedoms, health, economic welfare, social cohesion and environmental protection.

Matej Makarovič and Borut Rončević

## References

Archer, M. (2012). The Reflexive Imperative in Late Modernity. Cambridge: Cambridge University Press.

Archer, M. (2017). 'Introduction: Has a Morphogenetic Society Arrived?' In M. Archer (ed.), Morphogenesis and Human Flourishing, pp. 1–27. Cham: Springer International Publishing.

Asheim, T. B., Grillitsch, M. & Trippl, M. (2015). 'Regional Innovation Systems: Past – Presence – Future.' In: Doloreux, D., Shearmur, R. and Carrincazeaux, R. (ed.), Handbook on the Geography of Innovation. Cheltenham Glos: Edward Elgar.

Forray, D. (2015). Smart Specialisation. Opportunities and Challenges for Regional Innovation Policy. London: Routledge.

Fukuyama, F. (1989). 'The End of History', National Interest 16: 3–18.

Genov, N. (1997). 'Four Global Trends: Rise and Limitations' International Sociology, 12(4): 409–428.

Genov, N. (2018). Challenges of Individualization. Palgrave Macmillan.

Golob, T. (2017). 'Evropska študijska mobilnost kot sodobni obred prehoda', [European Study of Mobility as a Contemporary Rite of Passage], Glasnik Slovenskega etnološkega društva 57: 75–84.

Lasi, H., Fettke, P., Kemper, H., Feld, T. and Hoffman, M. (2014). 'Industry 4.0', Business & Information Systems Engineering, 6(4): 239–242.

Makarovic, M., Sustersic, J. and B. Roncevic (2014). 'Is Europe 2020 Set to Fail? The Cultural Political Economy of the EU Grand Strategy.' European Planning Studies, 22(3): 610–626.

# Part 1   Challenging Individuals

Mateja Rek

# Digital Gap: Media Habits of Youth and Senior Citizens in Slovenia

**Abstract:** In this contribution, we explore specific aspects of the media habits of youth and senior citizens (aged 65+) in Slovenia and identify which skills efforts to foster media literacy of both generations should focus on developing. The results of the analysis confirm the existence of a significant digital divide between the two generations. We describe how youth and senior citizens in Slovenia use media in very different ways. The critical issue with senior citizens' digital use is the large share of the older population that is anxious about trying to learn to use the internet, lacking confidence or being afraid of failure. Regarding youth, there are not many issues in access or abilities to use digital media. There are, however, shortcomings in their ability to reflect on media contents to which they are exposed and to participate in content creation processes in a personally and socially beneficial and desirable way.

**Keywords:** digitalization, media literacy, media education, youth, senior citizens, competences, critical assessment, training, lifelong learning

## Introduction

The public discourse on youth and their media habits is continuously marked by concerns about the effects of the characteristic of massive media exposure for contemporary youth, who may lack the maturity to enable them to make more informed choices or to recognize the positive and negative aspects of on-screen or online content. They are vulnerable to the dangers of developing unwanted behavioural patterns as a consequence of poor media choices. Concerns regarding this, as well as research interest, are often related to media violence and its effects on children and youth (Anderson et al., 2003; Bushman & Anderson, 2015; Kirsh, 2012), how often and to what ends youth use media and what effects media consumption have on their psychological development (Subrahmanyam & Šmahel, 2011; Calvert & Wilson, 2011), learning abilities (Dunkels et al., 2011), health issues (physical activity, obesity, sleeping habits) (WHO, 2010), self-perception and interpersonal relationships (Andsager & White, 2009; Kidd, 2018), the development of unwanted behaviours (such as drinking alcohol or smoking) (Anderson et al., 2009; White et al., 2015) and the like. Discussions on these issues are frequently accompanied by pessimistic forecasts.

The discourse on media habits of older people, in contrast, asserts that senior citizens lag in terms of engagement with digital technology (Friemel, 2016; Nixon, Rawal, Funk, 2016). Even though the number of older adults learning to use and engage with digital technology is growing (Eshet-Alkalai and Eran, 2009; Quan-Haase et al., 2016), a generational gap remains. The primary concern is the observed digital divide and the concerns that senior citizens have a disadvantaged position affecting their social engagement and participation in society as well as several aspects of quality living presupposed as a norm in contemporary society. Their non-use or low use of digital media is raising concerns about them being deprived of many services related to social security, well-being, health, networking, education, or personal development.

This paper aims to establish particularities in media habits of youth (aged sixteen to nineteen) and senior citizens (aged sixty-five and over) in Slovenia and to estimate how big the divide between these two generations in using digital media is. We aim to identify which skills efforts to foster media literacy of both generations should focus on developing.

In 2015, we conducted a survey entitled Media and Secondary School Students in Slovenia (Rek & Milanovski Brumat, 2016), gathering quantitative data on the media habits of secondary school students (aged fifteen to nineteen years) and analysed the impact of media habits in different areas of their lives. Data were collected using paper and online questionnaires; 818 secondary school students participated in the survey. Their participation was anonymous, not involving a name or any identifiable information about the subjects. Data were collected with the help of thirty-seven secondary schools, evenly located in all geographical regions in Slovenia, covering approximately the proportional distribution of rural and urban populations. In 2018, we conducted a similar survey, entitled Media habits of elderly people (aged 65+) in Slovenia (Rek, Kovačič and Brumat, 2018). In total, a sample of 322 individuals was assembled to represent the population studied as closely as possible. The survey was conducted in the presence of the assessor on paper or by typing directly into the computer. The data collection process was supported by the Slovenian Research Agency, in the framework of Infrastructure program of the Faculty of Media - Collecting, Managing and Archiving Data on Media Literacy. The questionnaires used in surveys differ as they were specifically designed to target particular circumstances of media engagement of different generational groups. However, some research questions addressing the media habits of both populations overlapped, offering us the possibility to identify generational particularities and divides. The results of the analysis confirm the existence of a significant digital divide between the two generations and describe how youth and senior citizens in Slovenia use media in very different ways.

## Characteristics of media habits of youth and senior citizens

As the use of digital technology can have significant benefits for seniors' quality of life (Bond, Burr, Wolf, & Feldt, 2010), many different stakeholders working with senior citizens, such as educators, civic actors, policy-makers, health specialists and scientists, are trying to find ways to support them in overcoming the barriers hindering their digital media engagement. Several barriers are hindering older people from using the internet and other digital devices, such as smartphones or tablets. Some may have difficulties with access to ICT devices or the internet, lacking money to afford them. However, studies show that in the population of older adults, not having skills to use digital devices and perform a wide range of online activities is a more significant barrier compared to the issue of access (Schreuers, Quan-Haase, Martin, 2017, Taipale et al., 2017). Reasons for the lower use of information and communication technologies and lower inclusion may also be related to health issues and disabilities (Bühler and Pelka, 2014; Ellis, 2016). Sight limitations can limit the perception of text and images on a screen, as well as coordinated eye-motor movements, such as selecting and validating online links with a mouse. Users with hearing problems do not distinguish between words and individual sounds, so they cannot detect audio alerts or spoken computer technology instructions; audio multimedia materials are thus inaccessible to them and, therefore, uninteresting. Cognitive and linguistic constraints include, for example, short-term memory problems or concentration problems. They can be very confusing, making it difficult for people to track navigation and complete on-screen tasks. With age, fine motor skills worsen, and people with reduced mobility may have difficulties in accessing a digital environment; however, with proper equipment and support, they can positively affect their quality of life.

There are also psychological barriers related to older peoples' digital engagement. Many are anxious about trying to learn to use the internet, lacking confidence or having a fear of failure. Digital literacy is gained through experience that goes beyond basic exposure to technology (Murray & Pérez, 2014), and older adults often lack the experience because they are hesitant to try new technologies (Quan-Haase, Martin, & Schreurs, 2014). Ageism can also have an impact on how an older person perceives their own ability to learn (Martínez-Alcalá, Claudia Isabel, et al., 2018). Older people reported less interest to learn in general as well as less interest in learning more about internet technologies and use compared to the active population, even though using doing so can have several benefits for their well-being.

The use of internet and mobile information and communication technologies can help older adults to communicate with their family and friends and stay in touch, especially with their grandchildren who are increasingly developing exclusively digital forms of everyday communication. Use of the internet can provide benefits in the form of access to information, be it on health issues, well-being, or possibilities for networking and life-long learning (Schreuers, Quan-Haase, Martin, 2017). However, the existing research suggests that adults may lack the necessary skills to critically evaluate the point of view from which information is presented (Livingstone, Van Couvering, Thumim, 2005).

As chronic loneliness and social isolation are significant concerns in the population of older adults, the development of digital skills and media literacy can contribute to the networking of older people and help them to find interesting forms of entertainment, joy and having fun.

In the generation of youth, who are 'digital natives' (Barlow, 1996) and grew up using digital devices, it is commonly assumed that they are more digitally literate than older people, who became acquainted with this relatively new technology at a later age. They can access, absorb, create and share media content more easily, even intuitively, without guidance or special training. This does not mean, however, that they can automatically the master media literacy skills needed to participate competently in contemporary communication flows. Media literacy is the ability of individuals to use and simultaneously autonomously and critically interpret the flow, content, values and consequences of the use of various media messages, as well as participate in the creation of media messages themselves (Martens, 2010; Martens & Hobbs, 2015). Understanding how media works in this changing environment and the ability to analyse and evaluate and how the reflexivity processes are related to various structural positions are crucial to media literacy (Livingstone, 2004; Golob and Makarovič, 2018; Golob and Makarovič, 2019). Mass media have become a significant agent in socialization processes and, at the same time, a new challenge for other socialization agents, such as family, schools or peer groups regarding their role in raising and educating children on the smart ways of using digital media, understanding content provided to them through mass media, making informed choices about media use and living with consequences of such choices. As children grow, the capacity of digitalization to shape their life experiences grows with them, offering plenty of opportunities to learn, to create, to connect, to play and to have fun. However, excessive, intense, reckless, or uninformed use of digital media can also have adverse effects on children and youth. Media-literate youth have better chances

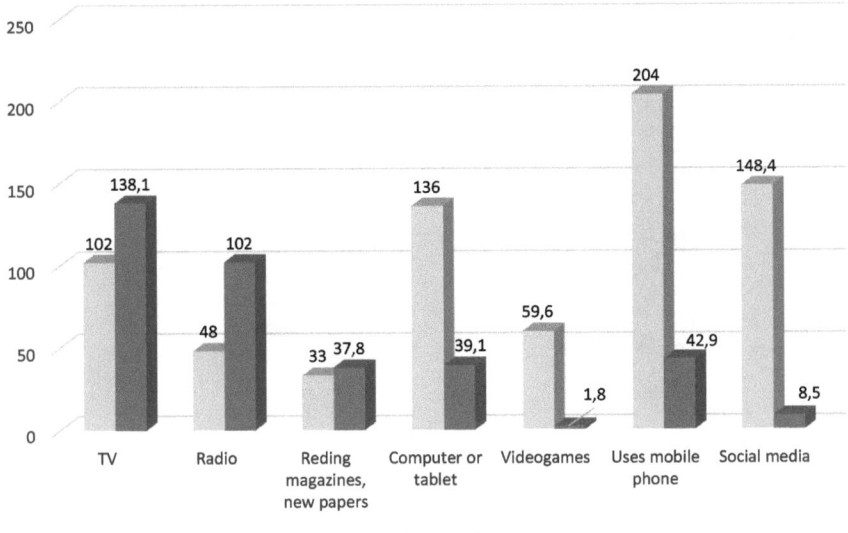

**Graph 1:** Media exposure of youth and seniors sixty-five and older (in minutes a day[1]).
Source: Rek and Milanovski Brumat, 2016; Rek, Kovačič and Brumat, 2018

to develop resilience to negative media phenomena, such as safety and privacy issues, hazards of virtual relationships and communication, hate speech, fake news or disinformation and similar. Educational systems are increasingly integrating media education in their curricula to empower children and youth in navigating digital media environments.

## How big is the gap in digital media use between youth and seniors in Slovenia?

The results of surveys on media habits of youth and seniors in Slovenia (Rek and Milanovski Brumat, 2016; Rek, Kovačič and Brumat, 2018) show that media are increasingly an integral part of their everyday lives. However, the choices and preferences in media use vary widely among both generations.

The senior respondents use more traditional media compared to the youth respondents. On average, they watch more TV and listen to radio more than twice as much as young people do. The habit of reading paper media is low in both populations. However, senior respondents spend far less time using computers, tablets or mobile phones compared to youth, as presented in Graph

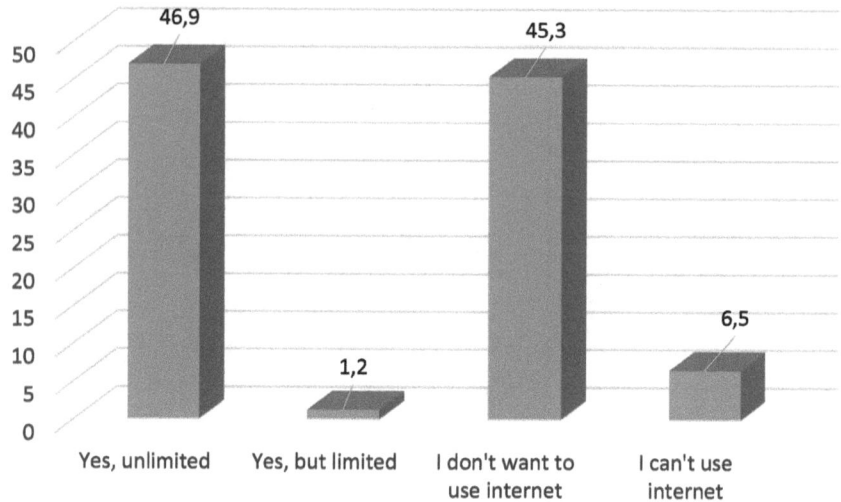

**Graph 2:** Do you have access to the internet? – Aged sixty-five and over. Source: Rek, Kovačič and Brumat, 2018

1. The difference is significant, especially in the case of the use of mobile phone and internet-related media content. More than half of the youth respondents use the internet between two and five hours a day, and only 10 % use it for one hour a day or less (Rek and Milanovski Brumat, 2016).

A total of 99 % of Slovenes aged sixteen to twenty-four use the internet, while only 47 % of Slovenes aged sixty-five to seventy-four years use it (Statistical Office of RS, 2018).

A significant share (45.3 %) of our senior respondents (Rek, Kovačič and Milanovski, 2018) claimed that they do not want to use the internet and did not opt to choose the option referring to their skills and ability to use it. Only 6.5 % of respondents chose the option 'I can't use the internet' when responding to the questions of the survey. Our findings support claims that the resistance of senior citizens to using digital technologies is not primarily related to actual ability issues, as widely believed. The values and beliefs elderly people have towards technology, wider concerns regarding its impact on society and fears of making mistakes (Knowles, Bran, and Vicki L. Hanson, 2018), anxiety and fear of failure (Murray and Pérez, 2014) are also significant factors holding back technology use among older adults. Even though the share of seniors using the internet daily has grown by 30 % in the previous ten years, as presented in Graph 3, a large

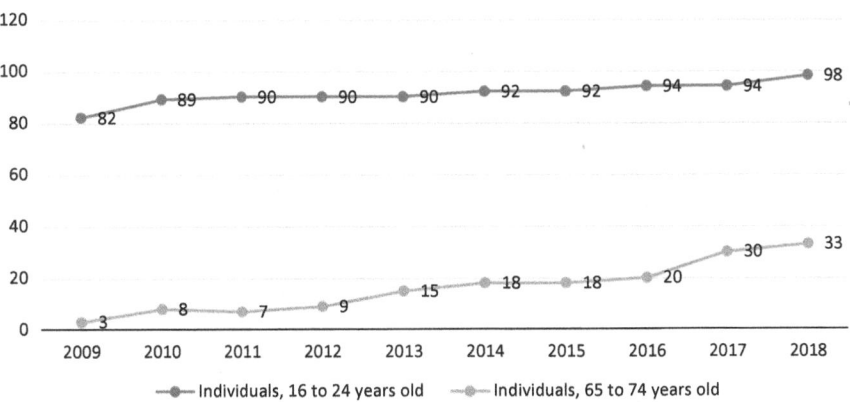

**Graph 3:** % of individuals using the internet daily. *Source: Eurostat database, 2019*

majority of seniors still does not use the internet regularly. There is a significant divide in daily presence online between youth and seniors in Slovenia.

That access to digital technology may not be the primary cause of non-use of internet and services internet-based technology among elderly people in Slovenia can also be seen from Graph 3. Only 1.2 % of respondents aged sixty-five and over do not use a mobile phone; 82.7 % of respondents own a smartphone. However, 52 % of those who do own a smartphone do not use the internet on it.

In both generations, the internet is used predominantly for communication purposes (sending/receiving e-mail, telephoning, or video calls), obtaining information (finding information about goods and services, reading online news sites/newspapers/news magazines), networking (participating in social networks) and especially entertainment (Eurostat, 2019). Activities related to online civic or political participation are among the least popular among both generational groups.

Among entertaining activities performed online, playing videogames is one in which only that generation is engaged; in the generation older than sixty-five, the activity of playing videogames is virtually absent. Youth spend, on average, an hour a day playing videogames (59.6 minutes). When choosing videogames, youth opt for violent ones in most cases, as the results of the survey show, that they, on average, spend almost three quarters of an hour (41.5 minutes), playing video games with violent content (beatings, shootings, etc.) (Rek and Milanovski Brumat, 2016).

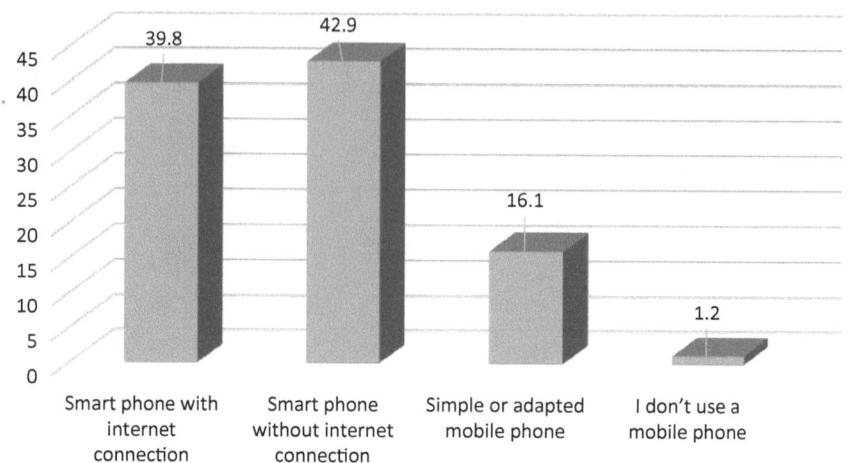

**Graph 4:** What kind of mobile phone do you use? – Aged sixty-five and over.
Source: Rek, Kovačič and Brumat, 2018

There are major differences in the use of social media between the two generations. From Graph 1 and Tab. 1 we can see that there is a significant gap in the use of social media both in the percentage of the population using social media and in the amount of time youth and senior users spend on social media daily. The average exposure time of youth respondents is two hours and twenty-eight minutes, and the average exposure time of senior respondents is far less (8.5 minutes a day). Only every fifth respondent aged sixty-five and older is present on social media, but only every tenth is present for half an hour or more daily (Rek, Kovačič and Brumat, 2018).

In our survey, respondents from both age groups were asked to provide details about the nature of the information they are posting online. They were asked whether they post personal information; information on what they have been doing lately; how they have been feeling lately; photos from trips, adventures and other experiences they have; photos of them in provocative poses (like inadequately dressed, with alcohol, drugs and similar); videos of other times that they have been caught doing something unusual or funny. They were also asked to assess their attitude towards posting various information: for instance, do they post information with pleasure; have some concerns but post information, or do not post specific information. Consistent with other research on generational characteristics of the use of social media (Chang et al., 2015; Kezer et al., 2016;

**Tab. 1:** For what do youth and seniors use the internet? Source: Eurostat data, 2019

| Activities on the internet - SI, 2018 | % of individuals 16 to 24 years old | % of individuals 65 to 74 years old |
|---|---|---|
| Sending/receiving e-mails | 93 | 37 |
| Telephoning or video calls | 63 | 14 |
| Participating in social networks (creating user profiles, posting messages or other contributions to Facebook, Twitter, etc.) | 89 | 14 |
| Finding information about goods and services | 86 | 37 |
| Internet banking | 35 | 16 |
| Selling goods or services | 20 | 6 |
| Making an appointment with a practitioner via a website | 14 | 8 |
| Seeking health information | 52 | 24 |
| Playing or downloading games | 46 | 7 |
| Listening to music (e.g., web radio, music streaming) | 88 | 16 |
| Watching internet-streamed TV or videos | 94 | 26 |
| Watching video content from sharing services | 88 | 23 |
| Playing/downloading games, listening to music or watching videos | 98 | 30 |
| **Activities on the internet - SI, 2017** | **% of individuals 16 to 24 years old** | **% of individuals 65 to 74 years old** |
| Uploading self-created content to any website to be shared | 67 | 5 |
| Reading online news sites/newspapers/news magazines | 74 | 28 |
| Travel and accommodation services | 44 | 15 |
| Job search or sending an application | 22 | 0 |
| Taking part in online consultations or voting to define civic or political issues (e.g. urban planning, signing a petition) | 6 | 1 |
| Posting opinions on civic or political issues via websites (e.g. blogs, social networks, etc.) | 8 | 1 |
| Civic or political participation | 11 | 2 |

Steijn, 2014; Van den Broeck et al., 2015), our findings indicate that older adults are less likely to create or post media messages on social media. They are less likely to disclose information, and they are less likely to disclose personal or provocative information either about themselves or others. Approximately half of the social media users in our survey among seniors are only viewers of media content: they do not create or post media content at all. Young people create and share information on their activities and feelings regularly, while the majority of them claim to reflect on the appropriateness of content posted on social media (Rek and Brumat Milanovski, 2016; Rek, Kovačič and Brumat, 2018).

## How big is the gap in the ability to critically assess media messages between youth and seniors in Slovenia?

The ability to reflect on media messages, to be able to critically analyse and evaluate such messages (received and created) and be able to form realistic responses to the complex ever-changing digital environment has become a necessity. Participants of online communication flows should be able to practice critical thinking, read or observe media texts and critically assess them, reflect on media texts in the framework of their own as well as broader social and cultural values. Media messages and representations are reductionist in comparison to reality. They can be simplified or distorted versions of reality or they can simply be wrong, fake, or even intentionally misleading. Particularly in the case of journalistic reporting, for which the guiding principle is the ideal of objectivity, or in the case of media messages, the purpose of which is to carry out an informative or interpretative function, the question of the accuracy of the messages and representations conveyed is crucial. The issue of trust in the accuracy of media messages transferred to the audiences has recently become highly emphasized. Given the spread of fake news and disinformation, especially in online environments, critical thinking and fact-checking have become a necessity in digital environments; this important function, aiming at sustaining the credibility of information shared in digital environments, can be performed by various stakeholders, including independent sources and fact-checkers from civil society or interested businesses as well as people, members of the audience themselves. The lack of accountability, quality and accuracy assurance tools of the messages conveyed through social media, blogs, sharing networks, social news and similar can lead to manipulative uses of communication infrastructures that have been harnessed to produce, circulate and amplify disinformation on a larger scale than previously, often in new ways that are still poorly mapped and understood (EC, 2018).

The authors of Trust in media in the EU report a noticeable trust gap between broadcast and new media. While radio and TV are the most trusted media among EU citizens, the internet and social networks are trusted the least. Trust in new media is at an all-time low, especially because of the rise of misinformation and disinformation online (EBU, 2019; Flash Eurobarometer 464, 2018). Data collected in our surveys among young people and senior citizens in Slovenia (Rek and Milanovski Brumat, 2016; Rek, Kovačič and Brumat, 2018) similarly indicates that there is a gap in trust between traditional and new media. Information provided by newspapers seems to be among more trusted compared to information provided on the internet. Compared to young people, senior citizens are more trusting towards content found on the internet and social media.

Given significant differences among both generations in their media use and the context of media use, we were also expecting a more significant deviation in the trust in media. For the majority of their lives, senior citizens in Slovenia were living in a totalitarian communist regime in which media messages were entirely under the political control of the communist party. In that time, critical thinking was not a desirable quality of an audience member. Obedience to propagandistic monolithic media messages was the audience behaviour expected by the system, and the principles of freedom of speech or plurality of thought were absent from society. Since 1991, when Slovenia became an independent, democratic state, the media landscape also started to develop in a more pluralistic manner, with respect for free speech, professional journalism and the free market.

Today's senior generation, which was raised and spent almost half of their lives living in a socialist system, had no training, education or capacity building on media in the transition to democracy. Young people did, and they never experienced the socialist media system. They are 'digital natives' born in a state of transition to democracy. Media education and critical thinking training have been an integral part of curricula at all levels of education in Slovenia since the 1990s. Additionally, there have been numerous programs and projects targeting children and youth offered by civic organizations aiming to foster pluralism, critical and out-of-the-box thinking. Given the generational differences in the societal and media contexts to which both generations have been exposed, we were assuming that the differences in trust in media would be more severe. We were also expecting that the younger generation, which did have media education as a part of their formal education curricula, would be more sensitive towards the issue of media ownership and financing because of their awareness of the possible impact of media ownership on media messages. They should be aware of the mechanisms of the functioning of the media industry and production.

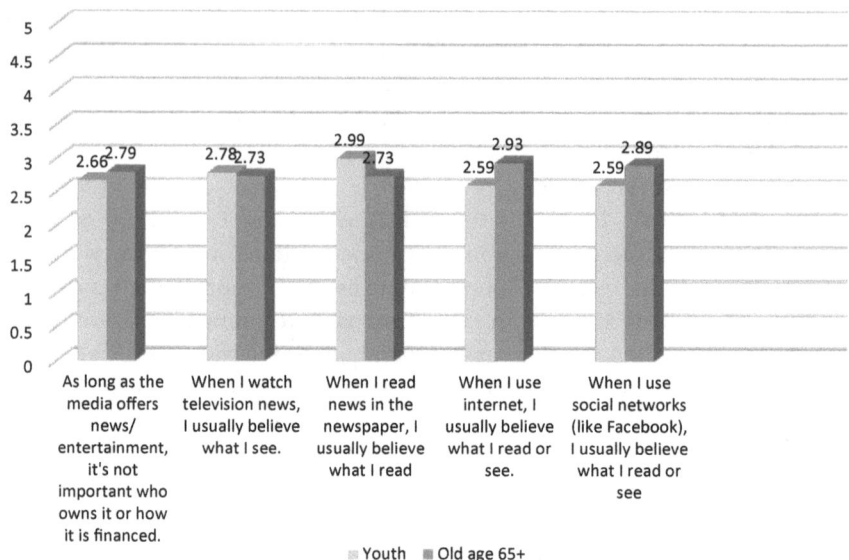

**Graph 5:** Trust in media – youth and seniors (five-point Likert scale: '1' means 'totally disagree' and '5' means 'totally agree'). Source: Rek and Milanovski Brumat, 2016; Rek, Kovačič and Brumat, 2018

However, as we can see from Graph 4, their attitude towards media ownership and financing is not that different from that of senior citizens.

## Conclusion

Senior citizens use more traditional media compared to young people. Conversely, senior citizens use far fewer online media compared to them. The habit of reading paper media is low in both populations. 90 % of young people use the internet for more than an hour a day. They are very active in performing various online activities, including playing videogames or spending time on social media, on which they spend 2.5 hours a day on average.

In contrast, 45.3 % of senior citizens do not want to use the internet. Only one-fifth of the surveyed population of senior citizens uses social media and those who do spend, on average, 8.5 minutes daily doing so. Among them, approximately half on social media users are only viewers of media content. They do not create or post media content.

In both generations, the internet is used predominantly for communication purposes, obtaining information, networking (participating in social networks) and especially entertainment (Eurostat, 2019). The percentage of individuals performing these activities online differs significantly among both populations. Activities related to online civic or political participation are among the least popular among both generational groups.

Both generations express more trust in traditional media compared to new media, especially in newspapers. However, reading paper media is the least popular form of media use among both populations. Senior respondents are more trusting towards content found on the internet and social media, compared to young people, even though they use it far less than they do. Critical assessments of the issue of the interrelatedness of media ownership and messaging distributed by media as well as their impact on audiences is similar in both generations.

When developing project and programs aimed at teaching skills in media literacy, they should be specifically targeted to the characteristics of the population for which they are intended. Given the large share of senior citizens that is not using new media, training and education efforts should continue to be mainly centred on the ability to use the internet and digital media. Attention should primarily be paid to overcoming anxieties and fears of related to the internet and new media use and persuading senior citizens to gain digital competences so a larger share of seniors will attend either training sessions teaching digital competence or become engaged in their forms of informal learning in their family of through peer-help.

The second, more challenging task, is to promote the concept of life-long learning, which is an idea into which the elderly population in Slovenia were not socialized. Younger generations in Slovenia have been systematically exposed to the idea of lifelong learning in the formal educational system. The competences needed to participate capably in digital environments are changing rapidly, as the environment itself is changing. People need to be aware that competences they gain in training or other ways will not benefit them forever. They need to stay in a learning mode permanently and continuously repeat the training and learning activities.

Given the large share of the elderly population that cannot or does not want to use the internet, it is only natural that the biggest concern in efforts to raise media literacy of this population is their ability to use new technology. Training aimed at raising their skills in critical thinking, their ability to analyse, critically assess media messages and be able to reflect on the state and developments in the media landscape is virtually absent. Even though senior citizens do not use new media to the extent younger people do, they still use media. They spend a significant amount of time using traditional media. Messages they are exposed to affect their way of thinking, behaving, perception of reality and they could benefit from training enhancing their ability to critically assess media messages and understand media landscape to which they are exposed daily.

Regarding young people, there are not many issues in access or abilities to use the internet. There are shortcomings in cognitive, emotional and social competences related to media messages or the ability to create and focus on creative problem solving, which are essential competences of a media-literate person. School curricula on all levels of education and other non-formal ways of teaching children and youth about media should not just teach children how to use it but should also include age-appropriate teaching that will help them develop skills to analyse media codes, to interpret and evaluate diverse media meanings and messages and to develop an understanding of the constructive nature of media messages. By including the explanations of how media work, children and young people should be (age appropriately) introduced to complex realms of cultural and social implications of mediated reality. Using digital media, young people can access many educational, creative, innovative resources as well as communication and networking activities that they would not have access to otherwise. Their task as media users is to learn how to become a smart and responsible consumer of digital products and how to participate in the content creation process in a personally and socially beneficial and desirable way.

# Bibliography

Anderson, C. A., Berkowitz, L., Donnerstein, E., Huesmann, R. L., Johnson, J. D., Linz, D., Malamuth, N. M., and Wartella, E. (2003). 'The Influence of Media Violence on Youth', Psychological Science in the Public Interest, 4 (3): 81–110.

Anderson, P., De Bruijn, A., Angus, K., Gordon, R., and Hastings, G. (2009). 'Impact of Alcohol Advertising and Media Exposure on Adolescent Alcohol Use: A Systematic Review of Longitudinal Studies', Alcohol and Alcoholism, 3 (3): 229–243.

Andsager, L. J. and White, H. A. (2009). Self-Versus Others. Media, Messages and the Third-Person Effect, Mahwah, New Jersey, London: Lawrence Erlbaum Associates Publishers.

Barlow, P. J. (1996). 'A Declaration of the Independence of Cyberspace', <http://editionshache.com/essais/pdf/barlow1.pdf >, accessed 3. August 2020.

Bond, G. E., Burr, R. L., Wolf, F. M., and Feldt, K. (2010). 'The Effects of a Web-Based Intervention on Psychosocial Well-Being among Adults Aged 60 and Older with Diabetes', The Diabetes Educator, 36 (3): 446–456.

Broadcasting Union (2019). 'Trust in Media', <https://www.ebu.ch/home>, accessed 2 October 2019.

Bushman, B. J., and Anderson, C. A. (2015). 'Understanding Causality in the Effects of Media Violence', American Behavioral Scientist, 59 (14), 1807–1821.

Bühler, C., and Pelka, B. (2014). 'Empowerment by Digital Media of People with Disabilities', International Conference on Computers for Handicapped Persons, Springer, Cham, 17–24.

Calvert, L. S. and Wilson, J. B. (2011). The Handbook of Children, Media and Development. Malden, Oxford, West Sussex: Willey-Blackwell.

Chang, P. F., Choi, Y. H., Bazarova, N. N., and Löckenhoff, C. E. (2015). 'Age Differences in Online Social Networking: Extending Socioemotional Selectivity Theory to Social Network Sites', Journal of Broadcasting & Electronic Media, 59 (2): 221–239.

Dunkels, E., Franberg, G. M., and Hallgren, C. (2011). Interactive Media Use and Youth: Learning, Knowledge Exchange and Behaviour. Hershey: IGI Global.

Ellis, K. (2016). Disability Media Work: Opportunities and Obstacles, Springer.

Eshet-Alkalai, Y., and Eran C. (2009). 'Changes Over Time in Digital Literacy', Cyberpsychology & Behaviour, 12 (6): 713–715.

European Commission (2018). The Multi-Dimensional Approach to Disinformation, Report of the Independent High Level Group on Fake News and Online Disinformation. Luxembourg: Publications Office of the European Union.

Eurostat. 'Eurostat Database. Digital Economy and Society Theme' (2019). <https://ec.europa.eu/eurostat/data/database>, accessed 27 September 2019.

Flash Eurobarometer 262. Fake News and Disinformation Online (2018). <file:///C:/Users/Mateja/Downloads/fl_464_sum_en.pdf>, accessed 16 September 2019.

Friemel, T. N. (2016). 'The Digital Divide has Grown Old: Determinants of a Digital Divide Among Seniors', New Media & Society, 18 (2): 313–331.

Golob, T., and Makarovič, M. (2018). 'Student Mobility and Transnational Social Ties as Factors of Reflexivity', Social Sciences, 7 (3): 1–18.

Golob, T., and Makarovič, M, (2019). 'Reflexivity and Structural Positions: The Effects of Generation, Gender and Education. Social Sciences, 8 (9): 1–23.

Kezer, M. et al. (2016). 'Age Differences in Privacy Attitudes, Literacy and Privacy Management on Facebook', Cyberpsychology: Journal of Psychosocial Research on Cyberspace, 10 (1).

Kidd, D. (2018). Pop Culture Freaks. Identity, Mass Media and Society. Second edition. New York and Abingdon: Routledge.

Kirsh, S. (2012). Children, Adolescents and Media Violence. A Critical Look at the Research. London, Los Angeles, New Delhi, Singapore, Washington: SAGE.

Knowles, B., and Vicki, L. H. (2018). 'The Wisdom of Older Technology (Non-) Users', Communications of the ACM, 61 (3): 72–77.

Livingstone, S. (2004). 'Media Literacy and the Challenge of New Information and Communication Technologies', The Communication Review, 7 (1): 3–14.

Livingstone, S., Van Couvering, E., and Thumim, N. (2005). 'Adult Media Literacy: A Review of the Research Literature', <https://bit.ly/39ROxMz >, accessed 3. August 2020.

Martens, H. (2010). 'Evaluating Media Literacy Education: Concepts, Theories and Future Direction', Journal of Media Literacy Education, 2 (1): 1–22.

Martens, H., and Hobbs, R. (2015). 'How Media Literacy Supports Civic Engagement in a Digital Age', Atlantic Journal of Communication, 23 (2): 120–137.

Martínez-Alcalá, C. I., et al. (2018). 'Digital Inclusion in Older Adults: A Comparison between Face-to-Face and Blended Digital Literacy Workshops', Frontiers in ICT 5: 21.

Murray, M. C., and Pérez, J. (2014). 'Unravelling the Digital Literacy Paradox: How Higher Education Fails at the Fourth Literacy', Issues in Informing Science and Information Technology, 11: 85.

Nixon, G. P., Rawal, R. F., and Andreas, F. (2016). Digital Media Usage Across the Life Course. London: Routledge.

Quan-Haase, A., Kim, M., and Kathleen, S. (2014). 'Not All on the Same Page: E-book Adoption and Technology Exploration by Seniors', Information Research: An International Electronic Journal, 19 (2): 2.

Quan-Haase, A., Martin, K., and Schreurs, K. (2016). 'Interviews with Digital Seniors: ICT Use in the Context of Everyday Life', Information, Communication & Society, 19 (5): 691–707.

Rek, M., and Milanovski, B. (2016). Mediji in srednješolci v Sloveniji (Media and Secondary School Students in Slovenia), Infrastructure Program on Media Literacy, Faculty of Media. <http://pismenost.si/mediji-in-srednjesolci-slovenija/>, accessed 27 September 2019.

Rek, M., Kovačič, A., and Brumat, K. (2018). Medijske navade starejših (65+) v Sloveniji (Media Habits of Seniors in Slovenia), Infrastructure Program on Media Literacy, Faculty of Media. <http://pismenost.si/pdf/Raziskava_Medijske_navade_starejsih_2018.pdf>, accessed 27 September 2019.

Schreuers, K., Quan-Haase, A., and Martin, K. (2017). 'Problematizing the Digital Literacy Paradox in the Context of Older Adults' ICT Use: Aging, Media Discourse, and Self-determination', Canadian Journal of Communication, 42 (2): 1.

Statistical Office of RS (2018). 'Development and Technology', <https://www.stat.si/StatWeb/en/Field/Index/25>, accessed 3. August 2020.

Steijn, W. M. (2014). 'A Developmental Perspective Regarding the Behaviour of Adolescents, Young Adults, and Adults on Social Network Sites', Cyberpsychology: Journal of Psychosocial Research on Cyberspace, 8 (2).

Subrahmanyam, K., and Šmahel, D. (2011). Digital Youth: The Role of Media in Development. New York, Dordrecht, Heidelberg, London: Springer.

Taipale, S., Wilska, T. A., and Gilleard, C. (eds.) (2017). Digital Technologies and Generational Identity: ICT Usage Across the Life Course. Abingdon and New York: Routledge.

Van den Broeck, E., Karolien, P., and Walrave, M. (2015). Older and Wiser? Facebook use, Privacy Concern, and Privacy Protection in the Life Stages of Emerging, Young, and Middle Adulthood. Social Media+ Society, 1 (2).

White. V., Williams, T., and Wakefield, M. (2015). 'Has the Introduction of Plain Packaging with Larger Graphic Health Warnings Changed Adolescents' Perceptions of Cigarette Packs and Brands?' Tobacco Control, 24 (2): 42–49.

World Health Organization Europe (2010). Protecting Children's Health in a Changing Environment. Report of the Fifth Ministerial Conference on Environment and Health. <http://pismenost.si/pdf/Raziskava_Medijske_navade_starejsih_2018.pdf>, accessed 19 September 2019.

Tea Golob and Matej Makarovič

# Does Digitalization Make Slovenian Youth More Reflexive?

**Abstract:** In our research, we follow the morphogenetic approach of M. Archer, who sees reflexivity as a mediator between structure and agency through individuals' inner dialogues. We apply a reflexivity measurement tool (RMT) as an approximate quantitative indicator of reflexivity levels and modes. The tool is extended to include the assessment of relational reflexivity as a potential for creating common (relational) goods. Using the data from an on-line survey among the Slovenian youth, we demonstrate that the technological environment has significant impact on their reflexivity. Using digital technology for public services, informal learning and following the news implies higher relational reflexivity, while its use for leisure and entertainment has the opposite effect.

**Keywords:** relational reflexivity, reflexivity measurement tool, youth, ICT, morphogenetic approach

## Introduction

In this chapter, we address the reflexive capacities of young people in Slovenia. The focus is on their abilities to adapt to contemporary social transformations while simultaneously also contributing to the morphogenesis of their society. The underlying concern herein is not only to what extent reflexive capacities contribute to more successful and empowered individuals but whether they are also linked to the emergence of a more cohesive and sustainable society.

Young people are called to be the future pillars of society. In a world moving so fast that we can hardly detect all the transformations underpinning our everyday reality, they are assumed to be the most capable of keeping pace with the compression of space and time. Living with technological advances from the cradle onwards, they are exposed to different cognitive, interactional and behavioural stimuli. They live in times when traditional social institutions, such as family relations and gender roles, have lost their rigidity and durability, while the future has become unpredictable and full of risks (Beck 1992). Tensions of individualization processes as a result of a much older historical trend (cf. Kleindienst 2019a; 2019b; Kleindienst and Tomšič 2018), strongly supported today by neoliberal market demands, offer new opportunities for individuals to become empowered, while they can also lead to disorientation and passive consumerism. Young people are seemed to be the most affected by pressures of constructing their biographies in terms of living individual life projects (Giddens 1991).

The key concept for understanding the role of individuals in these novel social settings is reflexivity, which stems from the recent sociological considerations attempting to understand the process of constructing the individual self in relation to the transformed social context. On the one hand, the term can be associated with the work of Giddens (1991) and Beck (1992), also offering a framework to a range of theorist such as Habermas, Lyotard, Bauman and Melucci (cf. Delanty 2000). On the other hand, there is an intellectual stream on reflexivity ensuing from the work of Margaret Archer (1988; 2003; 2007, 2012).

In our research, we follow the morphogenetic approach of Archer, who sees reflexivity as 'the regular exercise of the mental ability, shared by all normal people, to consider themselves in relation to their (social) contexts and vice versa' (Archer, 2007: 4). Through inner dialogue, they are able to define their concerns, develop projects and establish practices. Therefore, on the basis of reflexivity, individuals adopt certain 'stances' towards society, which constitute the micro-macro link and produce the 'active agent'. In that sense, reflexivity is a mediator between structure and agency (Archer 2003; 2007).

In her view, reflexivity is not just a regular activity of individuals, but it has become an imperative. Since social settings are no longer able to provide orientation in individuals' choices and practices, they feel an inner urge to be more reflexive. The generation particularly under pressure is youth, as social origins and socialization as such have become insufficient in providing proper grounds for social actions (Archer 2012).

However, it is not only the intensity of reflexivity to which attention should be paid but also the heterogeneity of inner dialogues recognized by Archer (2012). The dominant mode of contemporary society, specifically meta-reflexivity, calls for the relative autonomy of the structural and the cultural domains resulting in the morphogenetic society. Based on that, we are interested in the extent to which young people in Slovenia exercise such a mode of reflexivity and thus contribute to the morphogenetic processes.

When exploring the reflexive capacities of Slovenian youth, one should also explore the impact of the social environment in which individuals are embedded. As Archer (2003: 133) says, reflexivity refers to a nexus between a context contributed by the socio-cultural structure and concerns contributed by active agents.

The internal conversation is a response to the environment, which triggers emergent causal powers of individuals. In this chapter, we, therefore, attempt to identify how reflexivity is affected by individuals' positions in the social structure, such as those based on family background, education, gender and age. In that regard, we also take into account our previous research and other studies criticizing Archer's neglect of the role of the ascribed social origins and structural settings in one's reflexive deliberation (Atkinson 2010; Caetano 2015; Dyke et al. 2012; Farrugia and Woodman 2015; Mutch 2004; Mouzelis 2007; Porpora and Shumar 2010; Sayer 2010;).

However, it is not only the social environment that triggers our inner mental abilities, but it is increasingly also the technological one. Due to the rapid development of ICT, we can observe unprecedented access to information and knowledge. We are also able to establish new ways of interaction and participate in mobile and online communication. All those changes enable (and encourage) young people to spend more and more time online. In that regard, we are interested in how the different online activities of young people influence their intensity and also ways or modes of reflexivity.

Finally, when presuming that due to the contested social context and technological advances, young people are contributing to the morphogenesis, we also ask ourselves where those changes are heading? Is morphogenesis only referring to the further transformations of individualization trends encouraging empowered, self-referential, self-sufficient individuals (with the side effect of those who become disoriented and lost), or it can also encourage changes towards a more cohesive and sustainable society? In that regard, we focus on a special kind of reflexivity: the relational one, which is spreading beyond the micro-societal level and presumes the emergent properties and powers of relational subjects who, in their reflexive deliberations, take into account other concerns and share common goals with them.

By surveying Slovenian youth, we intend:

- To observe the presence of reflexivity levels and modes in our sample of Slovenian youth and compare it to our previous research on reflexivity,
- To upgrade our reflexivity measurement tool with an additional instrument to provide an approximation of relational reflexivity at an individual level,
- To determine, how family background, education, gender and age influence the reflexivity modes and the relational reflexivity,
- To determine which online activities influence the reflexivity modes and relational reflexivity,
- To discuss the consequences of our (preliminary) findings.

## Theoretical overview

Reflexivity as a concept is far from novel, as its roots can be traced back to Plato (Archer 2013). However, its empirical considerations are more recent scholarly endeavours. It has been initiated with Archer's attempt to sociologically ground the concept in theoretical terms inspired by the symbolic interactionists and pragmatist tradition. Based on critical realism, she rejects the ideas that the ontological domain of existence can be reduced to the epistemological domain of knowledge, which works against both positivist and constructivist ideas. She argues that there is an ontological subjectivity of every individual (2003). There is no structural determination directly influencing individual subjectivity. If the whole of reality can be divided into the social realm, the physical realm and the psychological realm, the first two have an objective ontology. In contrast, the psychological has a subjective ontology, meaning that 'objectively it exists, but subjectivity is its mode of existence' (Archer 2003: 38). The psychological systems have their own personal emergent properties, and reflexivity is one of them.

Her morphogenetic approach strongly opposes the stream of scholarly thinking that has been called 'extended reflexivity' (Adams 2006), which stems from interpretations of social transformations and increased individualization linked to the expansive changes in communication technologies and structures. It was explored by Anthony Giddens in his structuration theory (1984), which emphasizes the duality and dialectical interplay of agency and structure, and sees structural properties as both the medium and outcome of practices. On a basis of critical realism, Archer not only rejects his conflation between structure and agency being two sides of the 'same coin' but also points to the flawless of reflexivity as being a mere observation and monitoring of the continuing flow of activities and structural settings.

The problematic 'elision of structure with agency' (Archer 2013: 6), which is seen as fundamentally incompatible with reflexivity, Archer also finds in Ulrich Beck's (1992) reflexive modernization. Although Beck is mostly referring to systemic reflexivity and not social, his argument lies in the social pressure on individuals to become more reflexive. In line with Giddens, he argues that in late modernity, signified by individualization processes, people are not just encouraged but in a way are obliged to free themselves from the constraints of social structures. In that light, institutional positions determining individuals have started to present not just events and conditions influencing their lives, but at a minimum the consequences of the decisions that they make on their own (Beck 1992: 199). Although Archer similarly argues that the intense flow of transformation characterizing contemporary societies decreases contextual continuity and enhances reflexivity (2012; 2013), she says that 'increased reflexivity is not an automatic consequence or corollary of decreased routinization' (2013, p. 6). In her perspectives, reflexive deliberation always depends on a clear distinction of what is objective and what is subjective.

She has seen reflexivity as a process of inner dialogue changing through time and differing among individuals. By using a qualitative approach, she was able to recognize different modes of reflexivity, which preceded the quantitative tool ICONI, enabling her to determine consistent practitioners of each mode (Archer 2007; 2012). As Archer says, reflexivity takes place through inner dialogue, which is common to all people, but quite heterogeneous. Based on biographical interviews, she defined four different modes of reflexivity: communicative, autonomous, meta and fractured (see Archer 2003; 2007). Differences in modes exercised by individuals refer to a nexus between a context contributed by the socio-cultural structure, and concerns contributed by active agents:

Communicative reflexivity is defined by internal conversation, which needs to be confirmed and completed by others before they lead to action. The context is stable and continuous.

Autonomous reflexivity stems from the internal conversation, which is self-contained, leading directly to action and characterized by instrumental rationality. The initial context itself lacks stability.

Meta-reflexivity is based on the inner dialogue, which critically evaluates previous inner dialogues and is critical about effective action. It acquires an ultimate driving concern: to go no further than insisting upon relative autonomy of the structural and the cultural domains.

Fractured reflexivity stems from an internal conversation that cannot lead to purposeful courses of action and only intensifies personal distress and disorientation.

All those modes are practised in late modern society; however, there is a certain connection between the modes and social change. Archer (2003; 2007) has, for instance, argued that different periods induce particular modes of reflexivity. In traditional societies, the dominant mode of reflexivity is the communicative one, as it is collectivistic towards the social. Because of social transformations, uncertainties and 'contextual incongruity' between new openings and the expectations emanating from individuals' family backgrounds (Archer 2013, p. 9), one can see communicative reflexivity to be in decline, as young people are compelled to establish their own modus vivendi. In recent decades, new unpredictable and uncertain social areas have emerged that have influenced various transitions in everyday life. Modernity enabled autonomous reflexivity, which is accommodative towards social settings. Structural uncertainties have increased the importance of meta-reflexivity, which is transcendental towards the social, and also allow defining a sub-category of fractured reflexivity. She recognizes three categories of fractured reflexivity: displaced, impeded and near non-reflexive. Due to the mismatch between family backgrounds and novel life-opportunities, Archer argues (2013) that it is precisely meta-reflexivity that has become the predominant mode in the contemporary era.

For this chapter, it is crucial to differentiate between individual and collective meta-reflexivity (Archer and Donati 2015). Individual reflexive deliberations employ certain concerns and goals to be achieved, which can entail collective orientations, but only 'those attributable to singular persons and their aggregate effects' (Archer and Donati 2015, p. 61). Collective meta-reflexivity is about relational subjects, whose concerns are centred on the relation with the other. As has been argued, they are relationally constituted, and they generate emergent properties and powers through their social relations (Archer and Donati 2015, p. 31).

In that regard, we only exceed the level of individuals and move towards the structural social reality. Relational subjects imply social relations, which have their own properties and causal powers. It has been argued (Archer and Donati 2015, 55–56) that such relations are 'activity-dependent' and must be triggered by individuals, but have their own structure, whose causal powers work back upon the subjects. In terms of morphogenetic cycles of society, the active agents on the individual and collective levels are seen to be the ones who trigger emergent structural properties, which are responding with the new type of social differentiation, which is a relational one.

New structures that are emerging represent an upgrade to the modern society, which are 'constructed in such a way as to be immune to ethics' (Archer and Donati, 2015, p. 231). The prevailing systemic level based on the neo-liberal order does not encourage social solidarity and generates catastrophic consequences for the natural environment. Donati (2011) sees a viable alternative in the actions of relational subjects – generating common relational goods based on their collective meta-reflexive deliberations: relational reflexivity, which would lead to a new social order based on the relational differentiation and enables the social systems to go beyond the challenges and limits of the prevailing functional differentiation.

The idea has been adapted from Niklas Luhmann's social systems theory, which sees the primary mode of social differentiation as the key distinguishing feature between different archaic, traditional and modern societies. While the first one is characterized almost exclusively by segmentary differentiation (i.e., differentiation into similar units based on common descent or residence), the second is mostly based on stratificatory differentiation (based on distinction in rank). Modernization is thus seen by Luhmann as a shift from the previous social differentiation modes to functional differentiation. This is the differentiation than generates self-referential, self-organizing functional subsystems combining their strong interdependence with high levels of operational autonomy. This implies that (modern) society consists of relatively autonomous but mutually interdependent units that are functional subsystems (e.g., politics, the economy, religion, science, etc.), each based on its own specific principles and specialized in performing its specific function(s). While other modes of differentiation, such as divisions based on the nation and ethnicity (segmentary), extreme inequalities going beyond simple economic divisions (stratificatory) or between the global developed core and its periphery (core-periphery) do persist, the primacy of functional differentiation is becoming undisputable according to Luhmann (1999).

Since society and its subsystems as understood by Luhmann consist of communication, the central attributes that should be taken into consideration when applying Luhmann's perspective on social systems are the communication media that include language, transmission media and symbolically generated media (Luhmann 1999). The technology of transmission is crucial in this regard. The primacy of the stratificatory differentiation over the segmentary was thus enabled by written language, while the modern domination of functional differentiation was enabled by the invention of the printing press.

Donati (2011) considered relational differentiation to be a new type of social differentiation induced by the emergence of the World Wide Web and increasingly advanced technological development enabling new forms of communication to emerge. In a similar way as the printing press has enabled the creation of broad and comparatively inclusive national 'imagined communities' (Anderson 1991) connecting their members beyond direct presence and beyond straticatory divisions, the Internet has generated virtual communities (Rheingold 2000) based on online communication in all directions regardless of any distances in the physical space. Moreover, it has opened entirely unprecedented new ways of creating relational goods (Donati 2011) based on cooperation in the virtual space, such as the user-generated content of Wikipedia or open-source programming.

The concepts of relational reflexivity and relational differentiation also address another major challenge that seems to be unresolved, not only in Luhmann's theory but also in the existing empirical reality. While it has been able to generate unprecedented prosperity, the autonomous (market-based) logic of the economic subsystem has become increasingly destructive for the natural environment and social solidarity. Also trapped within their functional principles, the other functional subsystems also hardly seem capable of solving these problems. The operations of the political subsystem, for example, seem to be severely isolated in space (i.e., focused on the demands of the segmentary national constituencies) and time (i.e., planning until the next electoral cycle or, at best, within the leaders' and their voters' expected lifespans), since the functional logics of politics rewards nothing else.

Relational differentiation is supposed to go beyond the isolated logics of functional subsystems. It is manifested through a variety of intermediary associations or other ways of online or off-line collaboration, intended to generate common relational goods. Of course, for this purpose, it may interact with a variety of functional subsystems and their symbolically generated media. Providing relational goods may, for instance, imply activating power (politics), obtaining money (economy), searching for truth (science), referring to the legal rules (law), generating news (media) and similar, but it would still operate beyond the purely functional perspectives of these subsystems. One can hardly imagine a revitalization of solidarities or the solutions to environmental problems without activating people for the production of common relational goods.

Adapting Archer's (2003) and Luhmann's concepts, Donati (2011) connects macro-level social differentiation with micro-level reflexivity. Although all reflexivity modes can be found in all societies, communicative reflexivity is best suited for segmentary and stratificatory differentiation, as it contributes to the reproduction of the traditional order. In a similar way, autonomous reflexivity corresponds to the primacy of functional differentiation, in which highly individualized persons (not collectivities!) participate in a broad variety of functional subsystems as producers and consumers (economy), voters (politics), students (education), lovers (intimate relations) and similar. However, the intensive and diverse communication and mobility that thus take place in a modern functionally differentiated society (both because of its structural and cultural principles and because of growing technological possibilities), encourages the rise of meta-reflexivity, providing a critical stance both towards one's reflexivity and towards the existing social order. By extending meta-reflexivity from the individual to the collective level, relational reflexivity increases and, according to Donati (2011), becomes the basis of an emerging social order based on relational differentiation.

In our research, we intend to provide at least a rough insight into the presence of different reflexivity modes and particularly relational reflexivity among the youth in Slovenia. We see this as a way of assessing the potential for the generation of common relational goods and as an indication of how realistic it is to expect the rise of relational differentiation. Slovenia may be a compelling case to study in this regard as a part of the developed core when observed in global terms, but a (semi)periphery when observed in European terms.

Moreover, as the existing theories of reflexivity have only established some broad connections between information technology and reflexivity (Carrigan 2017; Donati, 2011), we need to observe this issue more precisely. This leads to our first research question: Which online activities, as the key part of the contemporary technological environment, contribute to different reflexivity modes and relational reflexivity of the Slovenian youth?

In addition, despite the morphogenetic social change, the existing research (Caetano 2015; Dyke et al. 2012; Farrugia and Woodman 2015, etc.) indicate that individuals' positions in social structure still matter for their reflexivity levels and modes. This may also imply that not all individuals are currently equally able to contribute to relational reflexivity. We will formulate this through our second research question: Do family background, education, gender and age as the relevant aspects of individuals' position in the social structure contribute to different reflexivity modes of Slovenian youth?

## Reflexivity measurement tool: towards the assessment of relational reflexivity

A reflexivity measurement tool (RMT) intended for indicating reflexivity levels and modes has been carefully developed and tested in terms of validity and reliability through our previous qualitative (Golob 2017) and quantitative (Golob and Makarovič 2018) research. In this chapter, we are extending our tool to incorporate an approximation of relational reflexivity. First, we will summarize our existing measurement tool for reflexivity levels and modes, following which we will present an additional way that may provide an approximation of relational reflexivity.

### Measuring reflexivity levels and modes

ICONI, Archer's original quantitative tool, consists of 13 items. The autonomous, meta and communicative modes of reflexivity are measured with three items each, whereas fractured reflexivity is measured with four items (detailed description in Archer 2007). Since there have been doubts about the validity (Meriton 2016) and internal reliability (Dyke et al. 2012) of the index, we upgraded the measurement tool.

We applied the exploratory-sequential model, in which qualitative research precedes the quantitative one, thus considering the limitations of quantitative deductive research. Our research tool in the questionnaire applied for the purposes of this research has thus been developed and tested through the following stages:

- a series of interviews with Slovenian students (presented in Golob 2017) as a qualitative approach to develop and test the validity of our subsequent quantitative instrument to measure reflexivity;
- a pilot application of the measurement tool on a convenience sample of Slovenian students;
- a pilot application of the reflexivity tool in a survey on mass media (administered online on a convenience sample) within the project Innovative Approaches of Encouraging Responsible and Pluralist Media in Slovenia;[1]
- an application of the tool on students from Slovenia (administered online on a representative sample of the Slovenian students engaged in Erasmus students' mobility and convenience samples of other students from Slovenia, Lebanon and the USA) (presented in Golob and Makarovič 2018; 2019).

This procedure has been consistent with the logic of critical realism that goes beyond both positivism and interpretivism (Alvesson and Sköldberg 2009; Wynn and Williams 2012). It is thus clearly based on methodological triangulation and not intended to provide any sweeping generalizations or determinisms. Instead, we simply intend to contribute to the search for the causal mechanisms that influence individuals' reflexivity, while being fully aware of the emergent properties of both the individual agents and the social structures.

Our reflexivity measurement tool consists of nine statements to which the respondents can reply on the Likert scales ranging from 0 (never) to 4 (all the time). Five of them are taken from the Internal Conversation Indicator (ICONI) (Archer 2007) upgraded by Porpora and Shumar (2010: 212), asking 'how often do you':

- plan your future;
- rehearse what you would say in an important conversation;
- imagine the best and worst consequences of a major decision;
- review a conversation that ended badly;
- clarify thoughts about some issue, person or problem.

---

1    Co-funded by the (1) Slovenian Ministry of Education, Science and Sport, (2) Public Scholarships, Development, Disabled and Subsistence Fund of the RS and (3) the European Social Fund.

These five statements have proven to be practical in obtaining reflexivity levels. In the procedure presented in our previous work (Golob and Makarovič 2018; 2019), we first apply these five items, as suggested by Porpora and Shumar, to calculate the reflexivity levels:[2]

Equation 1: Cronbach Alpha for the five items

$$R = r_1 + r_2 + r_3 + r_4 + r_5 \qquad (1)$$

where the values from $r_1$ to $r_5$ indicate the answers to each of the reflexivity items on the Likert scales ranging from 0 (never) to 4 (all the time) and R indicates the reflexivity level, which is an index ranging from 0 (no reflexivity) to 20 (maximum reflexivity).

Then, we use the four statements listed in Tab. 1 (second column) to determine each of the reflexivity modes. We should strictly distinguish between the Likert scores (L) for the statements from the questionnaire listed in Tab. 1 and the actual scores for the corresponding reflexivity modes (M). For instance, individuals can make decisions with or without the full agreement of others, but neither of these necessarily means such decisions are linked to reflexivity. Decisions may arise from impulses or traditional habits and not reflexivity. Consequently, the frequencies of behaviours listed in Tab. 1 should only be seen as indicators of different reflexivity modes when combined with the levels of reflexivity (our R reflexivity level index). For example, a person cannot be meta-reflexive without being reflexive at all. In contrast, the Likert scores attained from the question dealing with meta-reflexivity would multiply with the reflexivity levels: if a person indicates a particular meta orientation ($L_{met}$) higher then 0, this orientation will multiply with her/his overall reflexivity level (R): thus, generating the actual score for a given reflexivity mode ($M_{met}$).

Unlike that of Porpora and Shumar (2010), our measurement instruments do not make an arbitrary binary opposition between the reflexive and the non-reflexive, since we are dealing with Likert scales, not binary variables. We thus do not see people as divided between reflexive and non-reflexive but as being more and less reflexive. In addition, when compared to the original ICONI, our instrument enables us to distinguish between the intensity in terms of reflexivity level and the concurrent practising of the reflexivity modes within the inner dialogue,

---

2    Cronbach Alpha for the five items from our sample of youth is 0.80, making the index sufficiently reliable.

whose conflation has been problematized in previous research (Dyke et al. 2012; Meriton 2016).

## Assessing relational reflexivity

Strictly speaking, relational reflexivity as such is not a feature of an individual personality. As Archer says (2013), it is a collective reflexivity, which can be conceived in the same manner as modes of personal reflexivity. However, it does not imply collectivities in terms of sharing the same beliefs and intentions, but in terms of having a particular relation through which we strive for common commitments. Therefore, by its definition, it should be seen as an emergent property of a social relation. It thus exists in social relations, not in individual personalities. However, this does not prevent us from assessing its presence through the empirical observation of individuals' behaviours as reported through a social survey.[3] This becomes even clearer when we apply the concept of relational subjects – referring to those individuals that are engaged in the processes that contribute to the creation of relational goods (while trying to avoid relational evils) by contributing to relational reflexivity.

Although relational reflexivity can refer to all recognized modes of reflexivity, it is a meta-reflexivity that seems to contribute/trigger to the relational differentiation. It refers not to an individual but to relational entity intended to generate relational goods. This implies the presence of three elements:

1. Meta-reflexivity: One should distinguish between individual and collective meta-reflexivity (i.e., being meta-reflexive at the relational level in terms of critically reflecting on the reflexivity on how to generate relational goods and avoid generating relational evils). However, meta-reflexivity at the personal level is a necessary (though not by itself sufficient) precondition for relational reflexivity as a person cannot contribute to relational (meta-) reflexivity without herself or himself being meta-reflexive (cf. Archer and Donati 2015).

2. Activity in associations, clubs, NGOs and similar, which requires entering various relational entities that go beyond the most necessary relations based

---

3 This is comparable to the ways of assessing social capital according to those theories that see it as a social, not an individual property. This does not prevent them from observing social capital through membership in associations or generalized trust, reported by individual respondents in social surveys (cf. Adam and Rončević 2003).

**Tab. 1:** Statements indicating different reflexivity modes and the formulas for their calculation[1]

| Table 1 Reflexivity Mode | Mode of reflexivity indicator: five-level Likert scale each transformed into scores from 0 (never) to 4 (all the time): During the last year, how often did you... | Score based on the Likert scale (L) and its threshold | Formula for the calculation of the reflexivity mode (M) | Threshold for the reflexivity mode (M) |
|---|---|---|---|---|
| Communi-cative | Make important decisions with full agreement and support of the people close to you only. | $0 \leq L_{com} \leq 4$ | $M_{com} = L_{com} \times R$ | $0 \leq M_{com} \leq 80$ |
| Autonomous | Make important decisions based on your own best judgement regardless of what others think or say. | $0 \leq L_{aut} \leq 4$ | $M_{aut} = L_{aut} \times R$ | $0 \leq M_{aut} \leq 80$ |
| Meta | Carefully consider the key priorities of your life and why you are doing what you are doing. | $0 \leq L_{met} \leq 4$ | $M_{met} = L_{met} \times R$ | $0 \leq M_{met} \leq 80$ |
| Fractured | Feel lost and did not know at all what to do because of the things happening around you. | $0 \leq L_{fra} \leq 4$ | $M_{fra} = L_{fra} \times R$ | $0 \leq M_{fra} \leq 80$ |

[1] Adapted from: Golob and Makarovič 2018.

on employment, formal education, kinship and friendship. However, this activity is not sufficient by itself. Firstly, it should be linked to a certain level of meta-reflexivity. Furthermore, it should be clearly distinguished from the situations in which relations are only used by an individual to contribute to her/his own personal gains, not for the common relational goods.

3. Behaviour (not just intentions or values) intended to contribute to common social and environmental gains, which can be seen as the most relevant example of common relational goods. Again, this type of behaviour is only relevant for our measurement tool as far as it is (a) linked to meta-reflexivity

(for example, not just blindly following the pre-determined formulas of social and environmental sustainability provided by a dominant social semantics but critically reflecting them as well) and (b) relational in its nature, meaning linked directly to the action of other people through activities in associations, clubs, NGOs and similar.

Our reflexivity measurement tool has already provided us with the personal level meta-reflexivity (point 1). In our survey, the combination of points 2 and 3 can be obtained from the combination of the following survey questions: 'During the last year, how often did you. . .' (on the Likert scale ranging from 0 (never) to 5 (all the time):

- 'participate in activities in favour of greater social justice or helping people in need?'
- 'participate in activities in favour of environmental protection?'
- 'actively participate in an activity of a club, society, youth centre, association or similar?'

These three items cannot be combined into a fully reliable index (Cronbach alpha only equals 0.67). This is not surprising, since they may imply different dimensions since activities in favour of social/environmental protection do not necessarily take place through associations and vice versa: activities of different clubs, societies, associations may serve members' interests only without any contribution to the broader relational goods. Consequently, we apply principal component analysis to determine which component is relevant. From the three (orthogonal) principal components, the first one contributes 61 % to the common variance and has significant positive loadings for all three variables. The second component only has a significant positive loading for the active participation 'in an activity of a club, society, youth centre, association or similar' (contributing 24 %), while the third one is mostly focused on environmental protection (contributing the remaining 15 %). Based on our understanding explained above, only the first principal component is thus relevant as a contribution to relational reflexivity, as it combines collective agency (club, society, youth centre, association, etc.) with activities in favour of the creation of (social and environmentally relevant) relational goods. This component (denoted as $L_{rel}$) is presented on the scale from 0 to 1 and multiplied by the score for meta-reflexivity ($M_{met}$) to obtain our approximate measure for relational reflexivity. This indicator ranges (just like our reflexivity modes scores) from 0 (minimal relational reflexivity) to 80 (maximal relational reflexivity) as presented in Tab. 2.

**Tab. 2:** Statements indicating different reflexivity modes and the formulas for their calculation[1]

| Reflexivity | Score based on the first extracted principal component from the three activity items | Formula for the calculation of relational reflexivity ($R_{rel}$) | Threshold for relational reflexivity ($R_{rel}$) |
|---|---|---|---|
| Relational | $0 \leq L_{rel} \leq 1$ | $R_{rel} = L_{rel} \times M_{met}$ | $0 \leq R_{rel} \leq 80$ |

## Sampling and survey administration

The survey sample used in our research has been obtained through an online survey based on a combination of snowball and convenience sampling. The survey was administered between 30 October 2018 and 22 April 2019 using the 1ka web portal (1ka 2018). The obtained national sample consists of 650 young adults aged from 19 to 29, with their demographic structure presented in Tab. 3. Obviously, due to the sampling method and the structure of the sample, we cannot consider this to be a representative sample. However, due to its size and internal diversity, it can be seen as sufficient to test our methodological tool for relational reflexivity and to obtain at least some preliminary insights on the relationship between reflexivity, technology and position in social structure among Slovenian youth.

**Tab. 3:** Demographic structure of the sample[2]

| Demography | Category | % |
|---|---|---|
| Gender | Male | 35.0 |
| | Female | 65.0 |
| Age (mean = 23.62; std. dev. = 3.15) | 19–20 | 20.0 |
| | 21–23 | 29.7 |
| | 24–26 | 26.9 |
| | 27–29 | 23.3 |
| Education | Primary or less | 1.8 |
| | Vocational | 10.5 |
| | Secondary | 48.0 |
| | Tertiary | 39.7 |

[2] Statistical analyses presented in this paper have been conducted using Stata software (StataCorp 2015).

## Reflexivity modes and relational reflexivity among the Slovenian youth

First, we can present the mean reflexivity levels and scores in our online sample of Slovenian youth and compare it to the survey we conducted in March 2018 on the representative random sample of the general Slovenian population using the computer-aided telephone interviewing (CATI) method (Golob and Makarovič 2019). As can be seen from Tab. 4, the results of our online sample are quite consistent with the representative national sample from the survey conducted a few months previous, when we compare the respondents in the same age category, which implies that our online snowball-convenience sample does not depart much from a representative random sample in terms of assessing the responsibility levels and modes. It also speaks in favour of the validity of our existing RMT.

Moreover, we can also reconfirm the finding from our previous research (Golob and Makarovič 2019) that youth are, on average, more reflexive than the rest of the population. The mean scores for communicative and autonomous reflexivity modes are slightly lower in our online sample when compared to the same age group in the representative national sample (though still higher than for all of the age groups combined) national sample, while the fractured reflexivity seems to be higher in our online sample than in the representative sample. Nevertheless, the results are practically identical for the reflexivity levels and the meta-reflexivity scores. The latter is of particular significance, as

Tab. 4: Reflexivity levels and modes: comparing mean scores for two different surveys[1]

| Mean sample values | Online survey of the Slovenian youth, October 2018–April 2019 (19–29 years old) | Representative national phone survey, March 2018 (18–29 years old only) | Representative national phone survey, March 2018 (complete sample) |
|---|---|---|---|
| Reflexivity level (R) | 12.85 | 12.25 | 10.50 |
| Communicative ($M_{com}$) | 27.28 | 30.15 | 26.43 |
| Autonomous ($M_{aut}$) | 31.14 | 33.23 | 29.36 |
| Meta ($M_{met}$) | 36.26 | 36.26 | 30.43 |
| Fractured ($M_{fra}$) | 24.66 | 19.24 | 16.14 |
| Relational ($R_{rel}$) | 17.39 | / | / |

[1] Sources: Golob and Makarovič 2019; own survey and calculations.

the meta-reflexivity scores are used in our formula to calculate the relational reflexivity score (that was not available in our previous national representative sample), which speaks in favour or a relatively reliable assessment of relational reflexivity despite the obvious sampling limitations of our online survey.

In addition to the differences in the mean values between the youth and the general population in terms of reflexivity levels and modes (with the youth being more reflexive), there are differences in the overall frequency distributions. While the reflexivity levels and modes of the general population seem to be normally distributed (Golob and Makarovič 2019: 12–13), this is not the case for the youth. While the distribution of youth reflexivity levels is negatively skewed (-0.45), the skewness of relational reflexivity scores is highly positive (1.27), as most of our respondents tend to have very low scores for relational reflexivity. Their distribution is presented in Fig. 1. Low scores for relational reflexivity are not surprising given its definition as it implies not only significant meta-reflexivity but also its upgrading with social activity combined with broader social and environmental concerns.

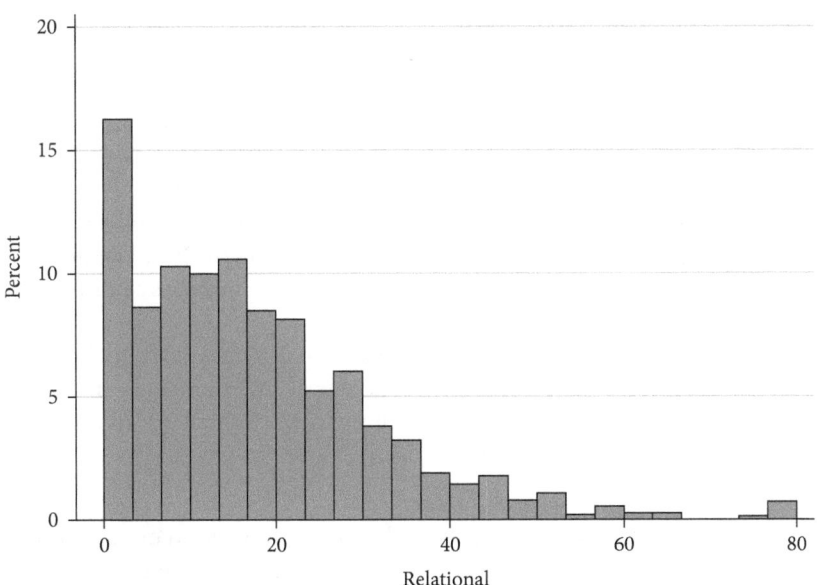

**Fig. 1:** The distribution of relational reflexivity scores

## Reflexivity in the context of embeddedness in social structure and the use of technology

Although reflexivity levels and modes are an individual's emergent property, they are triggered by the causal mechanisms of social structure and the individual's position and embeddedness in this structure. As Archer says (2003), reflexivity modes reflect the nexus between context and concerns. At the empirical level, this may be observed as a statistical relationship between reflexivity levels and different aspects of an individual's structural position, such as:

- gender,
- age,
- educational level,
- family background in terms of parents' education,
- material standard,
- inclusion in the study process and/or the labour market.

Obviously, these differences are typically not neutral: they may involve severe inequalities in terms of economic, cultural and social capital, as well as different types of formal and informal privileges or deprivations. Nevertheless, as indicated by previous research by (Adkins 2003; Farrugia and Woodman 2015; McNay 1999) and our findings (Golob and Makarovič 2019), the relationship between social inequalities and deprivations, on the one hand, and reflexivity levels and modes, on the other, are far from straightforward. While deprivations in terms of lower educational levels and lower material standards may imply lower reflexivity levels, the deprived position of women may make them even more reflexive than men. In addition, social deprivation and related frustration may become a major source of fractured reflexivity.

Nevertheless, people are not only affected by their social environments. They are also increasingly affected by their technological environments – through the ways how they are able to and how they chose to use technology. In our survey, we have considered the following aspects:

- Ordering products or services online,
- Using online public services, such as filing a tax declaration or applying for a visa online ('during the last year, how often did you. . .' for both items);
- Using digital technology for leisure and entertainment,
- Using digital technology for informal learning or following the news,
- Using digital technology for your job and/or formal education,

– Using digital technology to communicate with the people close to you (including social media, chats, phone calls, etc.) ('how much time per day do you spend on average for the following. . .' for the four items listed above).[4]

Since our reflexivity-related variables are not normally distributed, we have chosen not to rely on the classical parametric statistics but to apply a more robust measure, namely quantile regression analysis, which does not depend on the assumptions regarding the normal distribution. Using reflexivity levels and modes as dependent variables, we have built five quantile linear regression models taking into account that all reflexivity modes are obviously affected by reflexivity levels (as independent variables in these cases) and keeping only those structural and technological independent variables in the models that have turned out to be roughly statistically significant. The results provided by Stata software (Statacorp 2015) are presented in Tab. 5.

Finally, in Tab. 6, we provide a quantile linear regression model for the relational reflexivity potential. Again, we only include those independent variables that have turned out to be significant. Since the potential for relational reflexivity is not based on reflexivity levels but on the reflexivity modes, we have tested tentative effects of the reflexivity modes to relational reflexivity.

## Discussion

Contemporary society seems to be marked by unprecedented morphogenesis (Archer 2012). Macro-level configurations are providing an impetus for contextual discontinuity and incongruity, which have especially affected the young generations. The functional differentiation has offered new structural opportunities diminishing traditional social immobility, while at the same time the social semantic has become ambivalent, contested and hardly conceivable to individuals due to the immense flood of information. In addition, one can speak about structural and cultural asynchrony, which all together force people to re-evaluate their social positioning critically and also their attitudes towards the environment. Young people are, therefore, not just more reflexive but also reflexive in specific terms; that is meta-reflexivity. This mode of reflexivity is critical towards the social and towards the inner dialogues through which it also emerges.

Our study has shown that the reflexive tendencies of young people in Slovenia are consistent with Archer's (2012) findings, as the majority of the examined

---

4    On the impact of the technologies in the socialisation process, see also Rek 2019; Rek and Kovačič 2018.

**Tab. 5:** Quantile linear regression models for reflexivity levels and modes[1]

| Dependent variable / Independent variables | Reflexivity level (R) | Communicative reflexivity ($M_{com}$) | Autonomous reflexivity ($M_{aut}$) | Meta-Reflexivity ($M_{met}$) | Fractured reflexivity ($M_{fra}$) |
|---|---|---|---|---|---|
| Reflexivity level (R) | | **2** | **3** | **3.79** | **2** |
| Gender (Female) | **1** | | | | **9** |
| Age | **- 0.17** | | | | |
| Educational level (having tertiary education) | | | **7.5** | | |
| Material standard ("In material terms, my personal living standard is very good when compared to most of the other people in Slovenia." – level of agreement with the statement) | **0.33** | | 1.5 | | |
| Neither works nor studies | **- 3.17** | **- 5** | | **- 7.89** | |
| Ordering products or services online | | | | **- 0.30** | |
| Using digital technology for your job and/or formal education | | | | **0.30** | |
| Intercept | **15.5** | 0 | **- 13.5** | **- 11.62** | **- 9** |
| Variance explained by the model (Pseudo $R^2$) | **0.04** | **0.30** | **0.26** | **0.46** | **0.22** |

[1] Source: own survey and calculations; coefficients printed in **bold** are significant at less than 0.05 level; coefficients not printed are not significant.

population is indeed meta-reflexive. As Archer says, it is not an elision or demission of social structure that is reinforcing the reflexivity but the simultaneous acceleration of morphogenesis in social and cultural spheres. The relation between someone's context and concerns has turned into an impetus for individuals to become active agents and continually construct their primary and social

**Tab. 6:** Quantile linear regression models for relational reflexivity[1]

| Dependent variable | Relational reflexivity ($R_{rel}$) | |
|---|---|---|
| Independent variables | Coefficient | Significance |
| Communicative Reflexivity | 0.06 | 0.09 |
| Meta-Reflexivity | **0.46** | 0.00 |
| Educational level (having tertiary education) | 1.78 | 0.08 |
| Using online public services | **0.55** | 0.05 |
| Using digital technology for leisure and entertainment | **-0.95** | 0.04 |
| Using digital technology for informal learning or following the news | **1.13** | 0.02 |
| Intercept | -3.11 | 0.11 |
| Variance explained by the model (Pseudo R²) | **0.33** | |

[1] Source: own survey and calculations; coefficients printed in **bold** are significant at less than the 0.05 level; coefficients not printed are not significant.

identity within their morphogenetic cycles. However, despite globalising macro social constellations enabling morphogenesis to flourish, people have to activate structural emergent properties through their project first to enable their influence upon them. Youth in Slovenia is a social group substantially affected by individualization, also being considerably exposed to social risks and uncertainties (Ule 2008). However, there is no structural determination, and specific social embeddedness offering various life-chances is vital in that regard. Young people in Slovenia are exposed to variety in their social and technological environments, which influence their priorities and concerns, and consequently leading to different modes of reflexivity and agential abilities.

The differences within the young generation thus seem to be linked to the young individuals' structural positions. Young women tend to be more reflexive than young men, but they are also more fractured reflexive. This is consistent with our previous research dealing with the general population (Golob and Makarovič 2019) and the students (Golob and Makarovič 2018), as well as with previous research by other authors (Adkins 2003; Farrugia and Woodman 2015; McNay 1999).

For example, 'Women are supposed to compete for all social positions, just as men do, while they remain unable to escape from certain traditional gender-based limitations and expectations' (Golob and Makarovič 2018: 15). Like in the general Slovenian population, women are indeed triggered to express a higher level of reflexivity but are compelled to reduce their post-reflexive choices, which contributes to higher fractured reflexivity (Golob and Makarovič 2019). It is not just a process of reflexivity that is important but also 'what comes after' that moment of reflexive awareness (Adams 2006: 523). This is consistent with the argument that the reflexivity of women has been intensifying, but the remaining issue is how this leads to post-reflexive choices (cf. Adams 2006) in terms of actual reflexive deliberations (Golob and Makarovič 2019).

While having a tertiary education does not affect reflexivity levels, it does affect meta- and relational reflexivity. Having a certain educational level seems to make it easier for young people to develop critical attitudes and act accordingly, also in synergy with the others, accounting for the greater potentials of relational reflexivity.

As indicated in our previous research (Golob and Makarovič 2019), reflexivity levels are not only linked to social change in terms of intensified morphogenesis but also to individuals' personal morphogenesis. Even within the younger generation itself, younger individuals tend to be more reflexive.

One's perceived material standard also contributes to higher reflexivity levels and to autonomous reflexivity. It might be suggested that the lack of preoccupation with basic material concerns may enable people to become more reflexive. Perceptions of material welfare may also make people less dependent on others' support and thus contribute to higher autonomous reflexivity.

Finally, in structural terms, exclusion from both education and work seem to have devastating effects for reflexivity. Young people who neither work nor study thus tend to be less reflexive, less connected to the rest in terms of communicative reflexivity and less able to develop critical concerns regarding themselves and their social environment – in terms of meta-reflexivity.

While education seems to be the only aspect of individuals' structural positions directly affecting their potential for relational reflexivity, we should be aware of the crucial indirect effects of gender, self-perceived material standard and inclusion in the spheres of work and education, all of which affect relational reflexivity through reflexivity levels and meta-reflexivity.

A statistically significant relationship between relational reflexivity and meta-reflexivity can be seen simply as resulting from the fact that relational reflexivity, as we have presented it, is partly derived from individual-s meta-reflexivity. In contrast, though we can hardly draw any firm conclusions in this regard, we may note a positive relationship between communicative and relational reflexivity. This may imply – at least to some extent – that communicative reflexivity is not just a remnant from the past, characterized by more stable and predictable social relationships, but also a potential source of social connections and solidarity, relevant to some extend for relational reflexivity.

Moreover, it is not just the social but also the technological environment that has a significant impact on someone's reflexive deliberations. The technological environment turns out to be significant, especially when considering the potential for meta-reflexivity and – even more so – relational reflexivity. Our study shows that using digital technology for entertainment and leisure has a negative impact on relational reflexivity. In contrast, using digital technology for informal learning and following news has a strong positive impact on relational and meta-reflexivity.

On that basis, we can see that technology plays a vital role in enhancing or impeding the emergence of active agents and a morphogenetic society. At a tentative level, we can sketch three types of technology users in relation to their reflexivity modes. The first group is able to comprehend the complexity of internet information, are not digitally distracted by consumerist-leisure online activities (cf. Carrigan 2017) and are in emerging relations with others in order to produce common goods in terms of societal well-being. They typically use digital technology for informal learning, following the news and public services. They are highly meta-reflexive and have potentials for relational reflexivity. They are critically re-evaluating their social context, discerning their projects and dedicating to particular actions, which provide their well-being. They are, however, not only concerned with their well-being, but their ultimate concerns and modus-vivendi comply with generating common relational goods referring to the well-being of society. Due to their interest in active social participation, they might be called active digital citizens.

The second group is able to comprehend the complexity of internet information and is also not digitally distracted (Carrigan 2017). However, they do not share their ultimate concerns in emergent relations with others. Their use of technology is mostly related to their jobs and gaining formal education. They score highly in terms of meta-reflexivity but have no particular relationship towards relational reflexivity. Due to their instrumental relationship towards digital technologies, they might be called digital beneficiaries – benefiting from digital technology in practical terms related to their job and formal education.

The third group mostly seems to relate to the web content in a more passive way: as consumers. Extensive ordering of products and services online is related to lower scores for meta reflexivity, while extensive use of digital technology for leisure and entertainment implies significantly lower potentials for relational reflexivity. This group can be called passive digital consumers. They may be victims of 'digital distraction' in terms of being unable to cope with the 'cultural abundance' provided by the internet, which could lead to blindly following the social media (with their potentially limitless but actually quite limited variety) or the algorithms of commercial search engines (Archer 2017; Carrigan 2017).

For now, these relationships between reflexivity and technology can be seen as tentative ideal types. While we cannot categorically confirm them now, we can see them as a subject of further research in terms of validating their existence, determining their actual frequency and observing their behavioural specifics.

Aknowledgment: Part of this research has been co-funded by the Slovenian Research Agency (ARRS), project J5-1788 Advancing social and environmental sustainability: Exploring and predicting responsible behavior in Slovenia.

# References

Adam, F., and Rončević, B. (2003). 'Social capital: Recent Debates and Research Trends', Social science information 42, 155–183.

Adams, M. (2006). 'Hybridizing Habitus and Reflexivity: Towards an Understanding of Contemporary Identity?' Sociology 40, 511–528.

Adkins, L. (2003). 'Reflexivity: Freedom or Habit of Gender?' Theory, Culture & Society 20, 21–42.

Alvesson, M., and Sköldberg, K. (2009) Reflexive Methodology: New Vistas for Qualitative Research. London: SAGE Publications.

Anderson, B. (1991). Imagined Communities: Reflections on the Origin and Spread of Nationalism. London: Verso.

Archer, M. (1988). Culture and Agency. Cambridge and New York: Cambridge University Press.

Archer, M. (2003). Structure, Agency and the Internal Conversation. Cambridge and New York: Cambridge University Press.

Archer, M. (2007). Making Our Way through the World: Human Reflexivity and Social Mobility. Cambridge and New York: Cambridge University Press.

Archer, M. (2012). The Reflexive Imperative in Late Modernity. Cambridge: Cambridge University Press.

Archer, M. (2013). 'Reflexivity', Sociopedia.isa, p. 1–14. doi: 10.1177/205684601373.

Archer, M. (2017). 'Introduction: Has a Morphogenetic Society Arrived?' In M. Archer (ed.), Morphogenesis and Human Flourishing, pp. 1–27. Cham: Springer International Publishing.

Archer, M., and Donati, P. (2015). The Relational Subject. Cambridge: Cambridge University Press.

Atkinson, W. (2010). 'Phenomenological Additions to Bourdieusian Toolbox: Two Problems for Bourdieu, Two Solutions from Schutz', Sociological Theory 28: 1–19.

Beck, U. (1992). Risk Society: Towards a New Modernity. Thousand Oaks: Sage Publications.

Caetano, A. (2015). 'Defining Personal Reflexivity: A Critical Reading of Archer's Approach', European Journal of Social Theory 18: 60–75. doi:10.1177/1368431014549684.

Carrigan, M. (2017). 'Flourishing or Fragmenting Amidst Variety: And the Digitalization of Archive'. In Morphogenesis and Human Flourishing. In M. Archer (ed.), Morphogenesis and Human Flourishing, pp. 163–183. Cham: Springer International Publishing.

Delanty, G. (2000). Modernity and Postmodernity. London: Sage.

Donati, P. (2011). 'Modernization and Relational Reflexivity', International Review of Sociology—Revue Internationale de Sociologie, 21: 21–39.

Dyke, M., Johnston, B., and Fuller, A. (2012). 'Approaches to Reflexivity: Navigating Educational and Career Pathways', British Journal of Sociology of Education, 33: 831–848. doi:10.1080/01425692.2012.686895.

Farrugia, D., and Woodman, D. (2015). 'Ultimate Concerns in Late Modernity: Archer, Bourdieu and Reflexivity', The British Journal of Sociology. 66: 626–644.

Giddens, A. (1984). The Constitution of Society. Berkeley: University of California Press.

Giddens, A. (1991). Modernity and Self-Identity. Redwood City: Stanford University Press.

Golob, T. (2017). 'Evropska študijska mobilnost kot sodobni obred prehoda', [European Study Mobility as a Contemporary Rite of Passage], Glasnik Slovenskega etnološkega društva 57: 75–84.

Golob, T., and Makarovič, M. (2018). 'Student Mobility and Transnational Social Ties as Factors of Reflexivity', Social Sciences 7: 1–18.

Golob, T., and Makarovič, M. (2019). 'Responsible Behaviour of Youth', Proceedings of the 11th Slovenian Social Science Conference. Nova Gorica: School of Advanced Social Studies.

Kleindienst, P. (2019a). 'Pomen človekovega dostojanstva v delih Giovannija Pica della Mirandola [The Significance of Human Dignity in the Works of Giovanni Pico della Mirandola]', Ars & humanitas, 13 (1): 285–301.

Kleindienst, P. (2019b). 'Zgodovinski temelji sodobne paradigme človekovega dostojanstva [Historical Foundations of the Contemporary Paradigm of

Human Dignity]', Phainomena: glasilo Fenomenološkega društva v Ljubljani, 28 (108–109): 259–282.

Kleindienst, P., and Tomšič, M. (2018). 'Človekovo dostojanstvo kot del politične kulture v novih demokracijah: postkomunistična Slovenija [Human Dignity as a Part of Political Culture in New Democracies: Postcommunist Slovenia]', Bogoslovni vestnik: glasilo Teološke fakultete v Ljubljani, 78 (1): 159–172.

Luhmann, N. (1999). Die Gesellschaft der Gesellschaft. Frankfurt: Suhrkamp.

McNay, L. (1999). 'Gender, Habitus and the Field: Pierre Bourdieu and the Limits of Reflexivity', Theory, Culture and Society, 16: 95–117.

Meriton, R. F. (2016). Advancing a Morphogenetic Understanding of Organisational Behaviour: An Investigation into the Psychological Mechanisms and Organisational Behavioural Tendencies of Autonomous Reflexivity. PhD Thesis, University of Leeds, Leeds, UK.

Mouzelis, N. (2007). 'Habitus and Reflexivity: Restructuring Bourdieu's Theory of Practice', Sociological Research Online 12: 9.

Mutch, A. (2004). 'Constraints on the Internal Conversation: Margaret Archer and the Structural Shaping of Thought', Journal for the Theory of Social Behaviour 34: 429–445.

Porpora, D. V., and Shumar, W. (2010). Self-Talk and Self-Reflection: A View from the US. In M. Archer (ed.), Conversations about Reflexivity, pp. 206–220. London: Routledge.

Rek, M. (2019). 'Media Education in Slovene Preschools: A Review of Four Studies', CEPS Journal: Center for Educational Policy Studies Journal, 9 (1): 45–60.

Rek, M., and Kovačič A. (2018). 'Media and Preschool Children: The Role of Parents as Role Models and Educators', Medijske studije, 9 (18): 27–43.

Rheingold, H. (2000). The Virtual Community: Homesteading at the Electronic Frontier (2nd Edition). Cambridge: MIT Press.

Sayer, A. (2010). Reflexivity and habitus. In Conversations about Reflexivity. In M. Archer (ed.), pp. 108–123. London: Routledge.

StataCorp. 2015. Stata Statistical Software: Release 14. College Station: StataCorp LP.

Wynn, D., and Williams, C. K. (2012). 'Principles for Conducting Critical Realist Case Study Research in Information Systems', Management Information Systems Quarterly 36: 787–810.

Ule, M. (2008). 'Za Vedno Mladi? Socialna Psihologija Odraščanja', [Forever Young? Social Psychology of Adolescence], Ljubljana: Fakulteta za družbene vede.

Jana Krivec, Primož Rakovec and Tjaša Stepišnik Perdih

# The Role of ICT in Adolescents Dealing with Psychosocial Problems

**Abstract:** The search for psychosocial support is often stigmatized; therefore, mental health problems remain unresolved and may even worsen as a result. However, this stigma could be avoided by using modern information and communication technologies (ICT), which are an essential part of the lives of many young people. In this way, mental health care for young people would be increased, thereby preventively reducing the incidence of mental disorders in later periods and contributing to a healthier society. This paper examines how psychosocial problems can be tackled with ICT, the advantages and disadvantages of doing so, and some concrete examples of such support. The paper also presents an example of the development of a personalized online counsellor (OSVET) which combines virtual community, e-therapies and avatars; it is based on a cognitive-behavioural approach. OSVET has three phases: assessment, cognitive distortion identification, as well as personalized dialogue and appropriate assignments with an avatar. As such, OSVET could serve as a first step to help an adolescent cope with psychosocial problems.

**Keywords:** ICT, psychosocial support, adolescents, online counsellor

## Adolescents' psychosocial problems

According to Jeriček Klanšček et al. (2016), from 2008 to 2015, adolescents were most frequently treated due to stress reactions, anxiety disorders, hyperaesthetic disorders and eating disorders. As stated in the HBSC research report from 2014, almost half of young boys and girls have reported being overwhelmed with school and other stressful events. The same study also reports that 28 % of children and adolescents have at least two different psychosomatic symptoms per week due to various problems. The most common are insomnia, neurosis, irritation and melancholy (Jeriček Klanšček et al. 2015).

Adolescents experience stress for different reasons, most commonly in the academic, interpersonal, material and social domains (Anye et al. 2013; Hurst, Baranik, & Daniel 2013; Lee & Jang 2015). Students do not use stress coping strategies equally successfully. Some deal with stressors in a positive way (e.g., exercising, telling themselves that everything will be 'Okay'), whereas others start eating more, sleeping less, procrastinate and engage in other similar behaviour, leading to extended stress symptoms. Even though students are aware of the

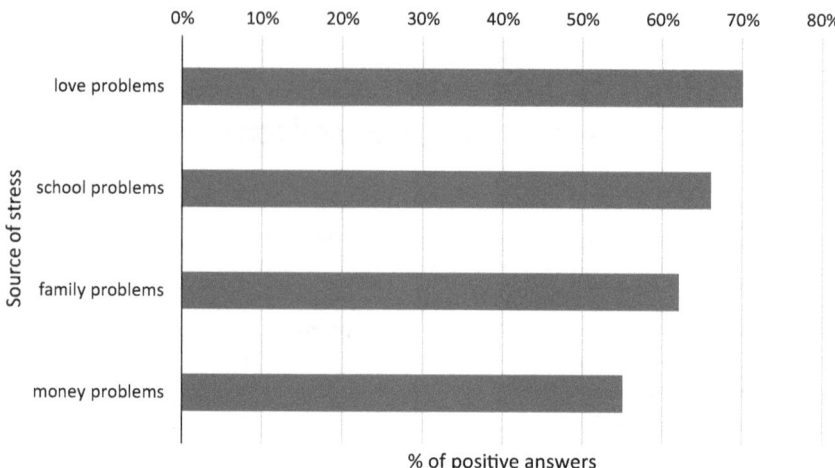

**Fig. 1:** Most frequent causes of stress among Slovenian students. Source: Adapted from 'Cognitive-behavioural profile of stress experience among Slovenian students' by J. Krivec, and P. Rakovec, 2018, Kairos, 12, p. 76. Copyright by Kairos

negative consequences of stress, they commonly attempt to solve the problems on their own, usually with unhealthy strategies and do not seek professional help (Baghurst & Kelley 2014; Dexter et al. 2018).

For that reason, we examined the incidence of stress among a selected sample of Slovenian students (N = 627) with detailed analysis of behaviour and thinking that they use when dealing with stressful situations. The results showed that the students 'often' experience stress. Only 6.68 % of participants said that they 'rarely or never' experience stress (Krivec & Rakovec 2018).

Most often, the causes of stress are relationship problems, problems at school, family problems and lack of money (see Fig. 1). Lack of interest, excessive expectations, public speaking, failure on exams and a sense of incompetence were also frequently mentioned. As well as that, they also cited stress stemming from health problems, thinking about the future, mental problems, problems with the law and residing in a student dormitory.

Students described several symptoms of stress, such as behavioural issues (sleep disorders, fatigue, aggression, acting impulsively, accelerated experience of time, difficult day-to-day functioning, no energy), physiological issues (difficulty breathing, stomach 'in knots', headache, cold sores, nausea, weight loss, chills, trembling), emotional issues (malaise, irritability, tension, restlessness, nervousness, intolerance, irritability) and cognitive issues (reduced concentration, focus on thinking about problems) (Krivec & Rakovec 2018).

The most frequent experience of stress is related to the behaviour of emotional relief (crying, shouting), lack of concentration, irritability and a sense of isolation. Results showed that over 30 % of participants in stressful situations repress their emotions (anger), which represents an avoided behaviour that can be considered as a form of indirect auto aggression. More than 50 % of participants experienced an exaggerated emotional state of stress and were overburdened with the result of a stressful situation (Krivec & Rakovec 2018). Such unhealthy behaviours derive from typical distorted ways of thinking, so-called cognitive distortions.

## Cognitive distortions and stress-related behaviours

How a stressful situation will be perceived and felt and what the behaviour primarily depends on the processing of information, which is often unconscious with the use of so-called 'automatic thoughts' (Beck 2011; Deal & Williams 1988; Kendall 1992). Machin and Creed (1999) define automatic thoughts as learned, unconscious thoughts one believes in and are difficult to control. The problem of automatic thoughts is that they are often dysfunctional or distortive, i.e., not real, in this case, they are called cognitive distortions.

The ability to deal with stressful situations is often influenced by typical cognitive distortions such as (1) mental filter (dwelling on the negatives and ignoring the positives), (2) disqualifying the positive (insisting that the user's accomplishments or positive qualities do not count), (3) overgeneralization (viewing one adverse event as never ending pattern of defeat), (4) catastrophizing (blowing things way out of proportion or decreasing their importance inappropriately), (5) personalization (blaming oneself for something one was not responsible for, or blaming other people and overlooking ways that the user's own attitudes and behaviour might contribute to a problem), (6) future telling (arbitrarily predicting things will turn out badly), (7) mind reading (the user assumes that people are reacting negatively to the user when there is no evidence for this), (8) shoulds and oughts (criticizing oneself or other people with 'should' or 'shouldn'ts', 'musts', 'oughts', 'have tos' and similar imperatives), (9) emotional reasoning (reasoning from how one feels), (10) global labelling (identification with one's shortcomings) and (11) low frustration tolerance (thinking one cannot and should not have to tolerate situations and conditions that are found frustrating) (Beck 2011).

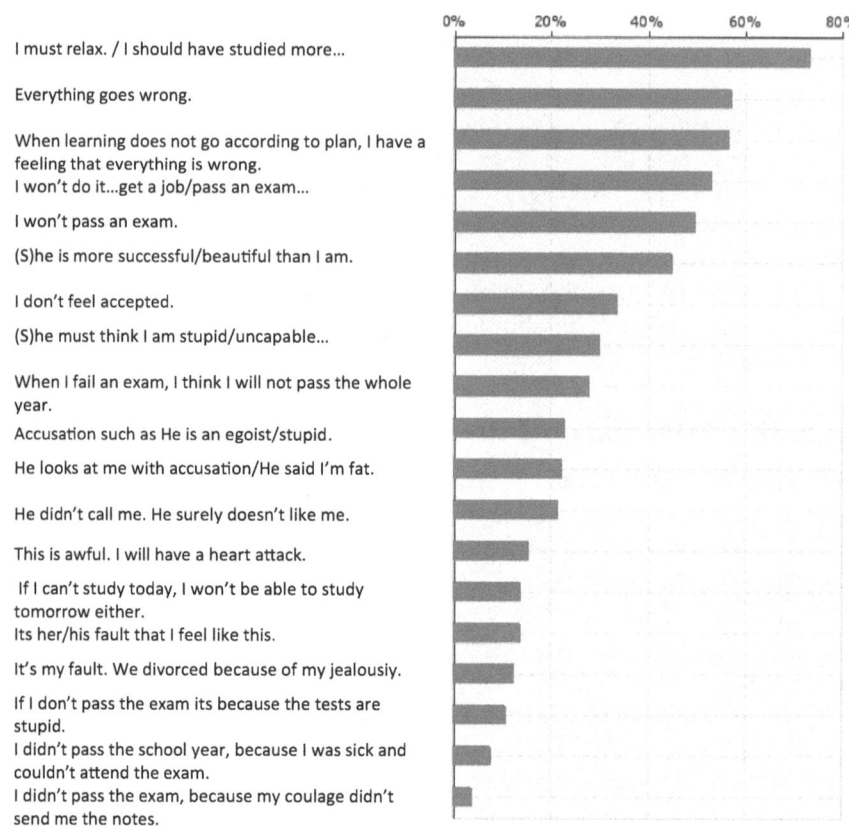

**Fig. 2:** Most common cognitive distortions of Slovenian students. Source: Adapted from 'Cognitive-behavioural profile of stress experience among Slovenian students' by J. Krivec, and P. Rakovec, 2018, Kairos, 12, p. 79. Copyright by Kairos

We conducted a study among 627 Slovenian quota-sampled students from different faculties (F = 481, M = 146) (Krivec & Rakovec 2018). Most of the respondents (76 %) were between twenty-one and forty years old, 20 % of them were younger than twenty years and 4 % older than forty. We have surveyed the occurrence of typical cognitive distortions among the selected students. The results showed that the most common cognitive distortion that occurs in the student's thinking is the formulation of 'should statements' (see Fig. 2 and Tab. 1).

We found a weak correlation between the frequency of stress and 'generalization' (r = .251 **, p = .000), 'all or nothing thinking' (r = .204 **, p = .002),

Tab. 1: Detailed descriptions of cognitive distortions

| Cognitive distortion | Detailed description |
| --- | --- |
| Low frustration tolerance | I have enough of all. I can't do it anymore. I would leave it altogether. Why do I need this?; Will I be able to? When will this end? Oh, these children. Why I do I exist? Why do we need to work and be unhappy? Constant struggle... |
| Catastrophizing | If I don't succeed, I don't know what the way forward is. When I make a mistake at work, it stresses me because I don't know what the consequences will be. |
| Future telling | I'll never finish. I won't find a job in time. I'm not gonna make it. I will not be able to pay major financial expenses. |
| Overgeneralization | When there are problems, everyone retreats. How can everything be so vicious?. I always have too little time. |
| Mental filter | Once again. School or job again. If I were more organized at least once. Why always me? |
| Labelling | I'm helpless. Inefficient. Egoist. |
| Personalization | Why should it work out for me? Look how many people are already doing that. Why do you think you should be something special? |
| All or nothing thinking | Everything will go wrong! |
| Shoulds and oughts | I must study again. |

Source: Adapted from 'Cognitive-behavioural profile of stress experience among Slovenian students' by J. Krivec, and P. Rakovec, 2018, Kairos, 12, p. 80. Copyright by Kairos.

'catastrophization' ($r = .238$ * *, $p = .000$), 'labelling' ($r = .208$ **, $p = .002$) and 'mental filter' ($r = .269$ **, $p = .010$). We have also verified the co-occurrence of the cognitive distortions in stressful situations. Results showed that 'mind-reading' is correlated with 'unjustified comparison' and 'mental filter'. 'Generalization' is correlated with 'should' statements. We have also used association rule learning (a priori algorithm), a rule-based machine learning method for discovering interesting relations between variables in large databases (Piatetsky-Shapiro 1991). Analysis with association rules showed that if there is a thinking pattern that includes 'generalization' and 'predicting the future' it is highly likely that 'should' statements will occur as well (support: 37, confidence: 0.94, lift:2.8) (Krivec & Rakovec 2018).

Correlations of coincidence of cognitive distortions and behaviour were analysed in three typical patterns of unsuccessful coping with the stressful situation, as proposed by Lazarus and Folkman's coping responses theory (1984): emotion-focused, cognitive appraisal and social support. As our study showed, each of these unsuccessful coping patterns includes typical connections of behaviours and ways of thinking.

1) When students were coping with a stressful situation with emotional relief, the following correlations between behaviour and thinking were discovered: behaviour focused on uncontrolled emotional relief crying ('In stressful situations, I would like to cry') is correlated with labelling ('He is a true egoist. I'm stupid. etc. ', r=.244, p=.000), unfair comparison ('(S)he is more successful than I am. (S)he is nicer than I am', r=.212, p=.000), catastrophization, all or nothing thinking ('Everything is wrong', r=.320, p=.000) and predicting the future ('I'm not going to. . . get a job/girlfriend/boyfriend/pass an exam', r=.244, p=.000). Emotional relief, which involves shouting, is, in contrast, correlated with mind-reading ('He certainly thinks ... that I'm lying to him / that I'm incompetent ... etc' (r=.201, p=.000), labelling ('He is a true egotist. I'm stupid. etc', r=.244, p=.000), mental filtering ('He is watching me again. He told me I was fat, etc', r=.306, p=.000), catastrophization ('This is awful. I will get a heart attack', r=.208, p=.000) and personalization ('It's my fault. We parted ways because of my jealousy', r=.228, p=.000).

2) Indecision is correlated with mind-reading, unfair comparison ('(S)he is more successful than I am. (S)he is nicer than I am', r=.275, p=.000), generalization ('When my studies are not going according to plan, I feel like everything is going wrong.' (r=.246, p=000), predicting the future ('I'm not going to. . . get a job/girlfriend/ boyfriend/ pass an exam', r=.250, p=.000), emotional judgment ('I feel they don't accept me', r=.200, p=000),) and mental filter ('He is watching me again. He told me I was fat, etc', r=.297, p=.000), Concentration problems correlate with mind reading ('He certainly thinks ... that I'm lying to him / that I'm incompetent ... etc' (r=.209, p=.000), unfair comparison ('(S)he is more successful than I am. (S)he is nicer than I am', r=.232, p=.000), generalization ('When my studies are not going according to plan, I feel like everything is going wrong.' (r=.276, p=000), predicting the future ('I'm not going to. . . get a job / girlfriend/ boyfriend/ pass an exam', r=.212, p=.000), mental filter ('He is watching me again. He told me I was fat, etc', r=.238, p=.000) and should statements ('I need to relax / I should learn more, etc', r=.269, p=000).

3) When considering lack of social support, we found that it is associated with generalization ('He didn't call me. That means he doesn't like me', r=.209, p=.000), predicting the future ('I'm not going to. . . get a job/girlfriend/ boyfriend/ pass an exam', r=.248, p=.000), all or nothing thinking ('Everything is

wrong', r=.217, p=.000), mental filtering ('He is watching me again. He told me I was fat, etc', r=.251, p=.000) and unfair comparisons ('(S)he is more successful than I am. (S)he is nicer than I am', r=.236, p=.000).

Our study also showed that students are aware of the negative consequences of experiencing stressful situations without adequate coping strategies. They expressed the need for some measures, including relaxation, professional help and conversation, which would enable them to deal with stress more effectively (Krivec & Rakovec 2018). However, the use of professional help and conversation remains stigmatized among Slovenians, even among the adolescent population.

According to Krivec and Suklan (2015), 94.8 % of Slovenians have experienced a state for which psychosocial professional help would be needed. If psychosocial issues are not treated properly and in time, serious psychological conditions and mental health disorders may result, which can affect daily lives. However, many of those in need do not receive the necessary treatment. One of the main reasons for this is the stigma surrounding it. Furthermore, in many cases we have relatively poor or unreliable information (see for example Makarovič et al, 2011). Studies (Krivec and Suklan 2015; U.S. Department of Health and Human Services, 1999) have revealed that nearly two-thirds of people who need mental health care never get it mainly because they are too embarrassed to make in-person contact with a psychotherapist who is part of the medical system. According to Krivec and Suklan (2015), 60.5 % of the interviewees think that psychotherapeutic practice should be widely accessible in social security or educational facilities, such as NGOs, schools or experts working in the psychosocial field and not incorporated exclusively into the medical system.

Stigma also affects the therapeutic process. Even if people find professional help, there is a significant discrepancy between the initial state and the number of people who successfully finish therapy. Moreover, people may not seek help due to its inaccessibility, especially if they live in a remote, rural area, far from a therapist's office. Next, scheduling, money, physical challenges, relationships marked by conflict, or misconceptions may also keep people from seeking help ('Metanoia: Online Therapy' n.d.). Finally, searching for therapy outside the medical sector can be problematic due to a vast number of unprofessional counsellors. Addis and Mahalik (2003) showed that strategies based on problems as a normal part of one's life enable an easier approach in seeking help to resolve them. Online psychosocial help is one of such strategies.

## The role of ICT in adolescents dealing with psychosocial problems

Information and communication technology (ICT) tools, such as text messaging, email, internet-based social networking and similar, have become a substantial part of adolescent life. Consequently, young people have been referred to as 'digital natives' in this 'bricolage of manifestations' in all spheres of life (Rončević and Tomšič 2017) as they communicate, seek information, engage, interact and entertain themselves through many technologies (Prensky 2012).

For that reason, ICT can be both, the part of the problem, such as ICT misuse, abuse, cyberbullying, causing psychosocial problems and an appropriate form to tackle psychosocial problems of adolescents (Kowalski et al. 2014; Livingstone et al. 2011).

Concerning psychosocial issues, for many, the Internet feels more private and helps them get past the barrier of stigma to seek help through e-therapy. In 2001, 80 % of Internet users or about ninety-three million Americans, searched for a health-related topic online (Best Counseling Degrees, no date). There are several different online facilities regarding psychological help. They can be classified in the following categories:

1. Virtual communities: self-help groups on the internet, where the psychosocial topics are widely spread; for example, AYAs or 7 Cups of Tea online support encourages members of the community to exchange emotional and informational support, coping with difficult emotions through expression (Love et al. 2012; '7 Cups' n.d.).
2. E-therapy and adjunct services: counselling platforms, where a large net of counsellors perform different kinds of therapies using virtual tools. Services are typically offered via email, real-time chat and video conferencing with professional psychologists in place of or in addition to face-to-face meetings (Mallen et al. 2005). Examples of good practice are BetterHelp ('BetterHelp' n.d.), Talkspace ('Talkspace' n.d.), Breakthrough ('Breakthrough: Confidential Online Counseling and Therapy' n.d.), Samaritans ('Samaritans' n.d.). The formation of the International Society for Mental Health Online (ISMHO) was a milestone in the development of e-therapy ('ISHMO' n.d.).
3. Computerized therapy virtual assistant: In this case, the computer is playing an active role in delivering clinical content. There are several ways of computer activities:

   a. Self-Help: Rules encapsulate clinical knowledge and are used to deliver targeted interventions. Adding more and more decision points and pieces of personalized content, the computerized 'self-help book' is becoming more extensive and more accurate. Eventually, each reader takes a unique path through the book based on their mental profile, which, in essence, is one of the key ideas behind computerized therapy (Helgadottir 2018). There are several such applications, such as Mood Tracker – a mobile application that allows users to self-monitor, track and reference their emotional experiences; PTSD Coach – provides opportunities to find support and tools that can help users manage the stresses of daily life with PTSD; LearnPanicCBT – self-treatment for panic disorder that is based on Cognitive Behavioural Therapy (CBT) principles; Stress Check – provides users with an overall stress score that illuminates their current level of stress; Cognitive Diary CBT Self-Help – helps to recognize thinking that interferes with achieving your goals in life and how to change that thinking.

b. Virtual reality therapies: use computer demonstration of reality most often to perform virtual exposure therapy for different kinds of anxieties and phobias, e.g., fear of flying. Examples of such applications are Virtually Better, Psious, VirtualRet, Mimerse (Senson 2016).

c. Robots: they are used for human-computer interaction, such as Paro, a therapeutic robot in the form of a baby harp seal developed by The National Institute of Advanced Industrial Science and Technology in Japan to comfort people with dementia to increase their motivation, reduce the stress of the patients and their caregivers and stimulates interaction between patients and caregivers ('PARO Therapeutic Robot' n.d.).

d. Avatars: they are based on human-computer communication in natural language. ELIZA is one of the earliest and most well-known programs that attempted to act as a therapist and provide Rogerian psychotherapy. Because Rogerian psychotherapy primarily encourages clients to talk more rather than engaging in a discussion, ELIZA takes the users text and rephrases it, putting the focus back on the user and encouraging him/her to talk more rather than conversing. This approach works for some clients but quickly becomes frustrating and pointless. It might be worth mentioning that the version of ELIZA developed in MIT's Computer Science Department, was among the most frequently visited in the world. The reason for this was found years later – it mixed replies from users online at the same time. Unfortunately, it was treated as non-privacy-preserving and consequently eliminated. Unlike ELIZA, a newer chatbot, Ellie, was able to talk about herself and generate a conversation rather than only rephrasing the responses it received. Ellie's aim was to treat people with depression and veterans with post-traumatic stress disorder (PTSD). The program recognizes facial expressions and analyzes audio and posture to formulate its response and adjust its tone. Nonetheless, Ellie was far from being able to provide the kind of understanding a human therapist could (Lucas et al. 2014). More recent applications are MoodKit ('CBT Apps' n.d.) that help the user identify and change unhealthy thoughts and chart the user's state of mind and Woebot ('Woebot' n.d.) which uses CBT to reduce depression. A virtual coach, 'Shelley', incorporates patient education materials and uses conversations to help the user make decisions and change behaviour.

## Efficacy and benefits of online psychosocial help

A growing body of research on online counselling has established the efficacy of online therapy with treatment outcomes being similar to traditional in-office settings (Mallen et al. 2005). Studies (Andersson et al. 2014; Cohen & Kerr 1999; Fitzpatrick, Darcy, & Vierhile 2017; Suler 2000; Wagner, Horn, & Maercker 2014) show that online psychotherapy yields the same or even better results than conventional therapy in the medium term (Rus-Calafell et al. 2013).

Rus-Calafell and colleagues (2013) confirmed the efficacy of using virtual reality in psychotherapy. Client satisfaction surveys also tend to demonstrate a high level of client satisfaction with online counselling (Dongier et al. 1986). In a recent survey of over 400 clients of online therapists, more than 90 % responded that working with a therapist on the Internet helped them ('Metanoia: Online Therapy' n.d.).

An analysis of Slovenian online counselling 'Tosemjaz.com' from 2014 determined that the internet is an instrument that allows informational, counselling and psychosocial support for adolescents in different crisis situations, such as self-harm and suicidal thoughts (Lekič et al. 2014). The most important benefits of online psychosocial support are: (1) it allows the client to attend sessions at a higher rate than traditional sessions and reduces the number of missed appointments (Glueckauf et al. 2002); (2) it is good for clients located in areas under-served by traditional counsellors (such as rural areas) or for disabled people, who traditionally under-utilize clinical services (Mallen et al. 2005) or clients who may have difficulty reaching appointments during normal business hours (Chang, Yeh, & Krumboltz 2001); (3) counselling in person is more likely after using e-counselling, 64 % of the persons moved on from e-counselling to consult a counsellor in person (Chang, Yeh, & Krumboltz 2001); (4) it is beneficial

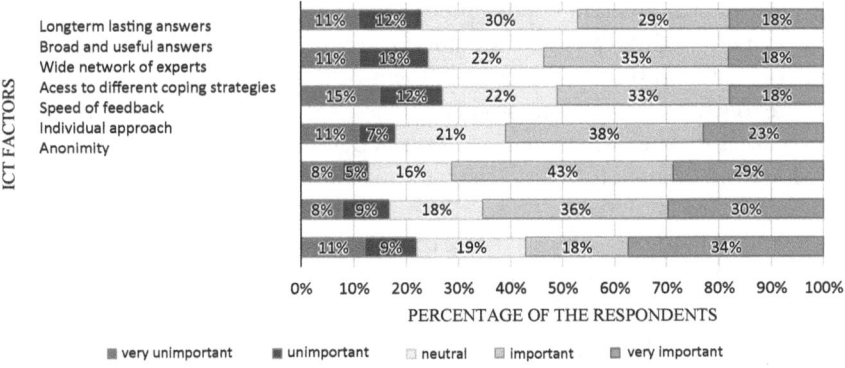

**Fig. 3:** Importance of ICT factors for solving adolescent's psychosocial problems. Source: Adapted from 'Positive impacts of information and communication technology in coping with adolescents with psychosocial problems' by J. Jemec, 2019, p. 55. Copyright by SASS

for young people, who are keen on using computers or smartphones; (5) it cheaper and thus more affordable; (6) it reduces social stigma; (7) anonymity encourages people to disclose their problems, emotions, thoughts or sensitive information, clients admit that they feel less judged by the virtual counsellor and more open to him/her, especially if they were told that (s)he was operated automatically rather than by a remote person (Lucas et al. 2014); (8) it enhanced content: computerized counselling can incorporate more than text on a screen. The programs can be rich in multimedia content, with images, videos, animations, audio voiceovers and interactive exercises; a well-designed treatment program can be a highly compelling user experience; (9) virtual reality is a protected environment for the client, where one can deal with a feared situation in secrecy.

## Dis/advantages of ICT technologies

According to Jemec (2019), most Slovenian adolescents rely on ICT for solving their psychosocial problems. Usually, adolescents first attempt to find information on the web and only then seek it in-person in conversation with friends, parents or counsellors. The most critical factor for that is anonymity. Other factors and their importance are presented in Fig. 3.

For adolescents, ICT represents a nearly infinite source of information that they can access at any time. They like that ICT provides various possibilities for problem-solving, as well as the possibility of contacting various professionals who can help them at different opportunities. When looking for the right answers, it is also vital that they can get them quickly. Writing, reading and watching content connected to their problems give them insight into a different world than the one to which they are accustomed. Indeed, they see that others have similar problems as they do, which makes them feel that they are not alone.

Moreover, through ICT, they overcome distance because it allows them to talk to people living on the other side of the world or with friends who are not around them at that moment. They are aware of opportunities that instant access to the web offers them, for example, YouTube music which helps them to forget about their problems even if only for a few moments. They also acknowledge the positive influence of the video content offered by their role models or by other individuals who, through their inspiring speeches and positive content, encourage and motivate them (Jemec 2019). Powell et al. (2010) conclude that ICT has the potential to (1) engage young people through media that are familiar to them; (2) support digital public services; (3) facilitate greater involvement of informed service users; (4) reduce reliance on 'real world' resources and interactions.

In contrast, there are also some disadvantages that adolescents perceive when using ICT for psychosocial problems. While they welcome technology-based services, they want these to be a complimentary provision rather than a replacement for face-to-face approaches (Powell et al. 2010). Jemec (2019) also determined that what bothers them the most is the lack of nonverbal communication. Seeing the direct response of the person with whom they are talking is crucial for them. In difficult times, furthermore, they also need genuine physical contact.

Another problem is the plethora of unverified and unprofessional information on the web and in applications. Adolescents do not trust the answers they get from people on the web because they doubt their professionality. This shows the high level of reflexive and critical attitudes of adolescents, but also reflects the current problematic situation of the lack of professional online systems for the provision of psychosocial help. Moreover, adolescents miss a more individual approach because, in their view, everything they can read, listen to or see through ICT is too general. They are also disturbed by the overwhelming amount of information that makes it difficult to determine which information is right or would work well for them. Some people even misuse the anonymity of such systems for criticizing and ridiculing others. Adolescents are aware that to be able to use different technologies, some basic technical conditions must be fulfilled, which is not always possible. They are also concerned about exposure and lack of anonymity, the theft of personal data and the permanent preservation of data on the web (Jemec 2019).

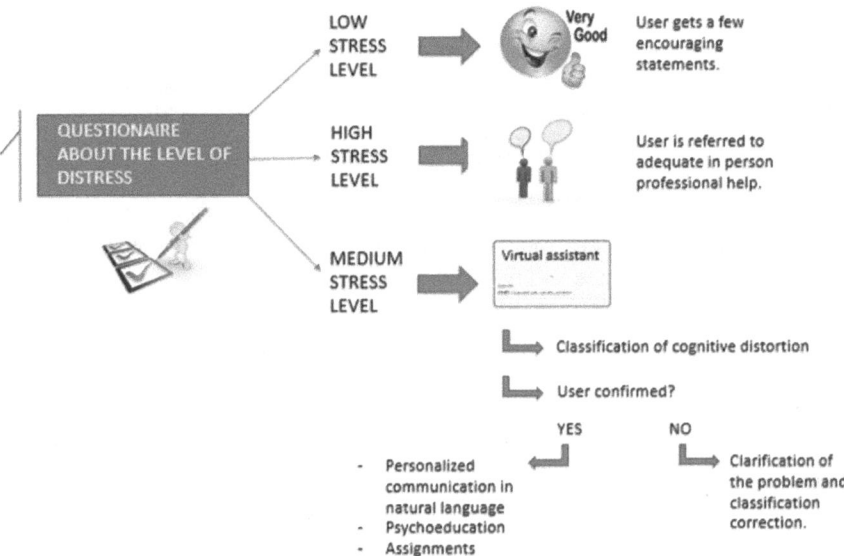

**Fig. 4:** OSVET architecture. Source: Adapted from 'Positive impacts of information and communication technology in coping with adolescents with psychosocial problems' by J. Krivec and M. Gams, 2017, p. 40. Copyright by Cognitonics: proceedings of the 20th International Multiconference Information Society - IS 2017, 9th–13th October

According to the presented dis/advantages that adolescents find in the usage of ICT for dealing with psychosocial problems, we propose an example of the development of a personalized online counsellor, called OSVET (the name is an abbreviation for the 'online counselling' in the Slovenian language and was the title of the project applied and approved by the Public Scholarship, Development, Disability and Maintenance Fund of the Republic of Slovenia, Slovenian Ministry of Education, Science and Sport, and the European Social Fund).

## A model of personalized online counsellor – OSVET

The scope of our solution, called 'OSVET', is to help people in distress. It combines virtual community, e-therapies and avatars. It includes the screening of the distress severity level, followed by personalized self-help and psychoeducation (see Fig. 4). The human-computer interaction (HCI) is performed in natural language, which gives the user a feeling as if she/he is interacting with a real person. In case of severe mental health issues, users are encouraged to proceed to e-therapy or seek in-person therapy, and appropriate contacts are provided to the user. With such an approach, we believe that stigma will be reduced, and people in need will have a pleasant experience searching for help, which will further encourage them to increase taking care of their mental health.

OSVET was developed on the following assumptions (Krivec & Gams 2017):

1. Combination of several therapeutic techniques: OSVET uses screening of the user state, psychoeducation, self-help, communication with virtual assistant using natural language and implements e-therapy if necessary.
2. Personalization: based on the information provided by the user, OSVET provides personalized answers, instructions and assignments.
3. User-friendly HCI: in addition to natural language conversation, the user experience is enriched with information about stress overload issues, cognitive distortions, exercises, etc.
4. Safety: OSVET is based on personalized psychoeducation, which is safe and positive in nature. When a cognitive error is classified by OSVET, the user is asked to confirm the assumption. If there is no confirmation, OSVET asks for clarification before moving to further steps. If a higher level of stress or more serious mental issue is detected, users are redirected to in-person therapists. Finally, the content of OSVET is designed by experts in the field of psychosocial counselling.
5. Cognitive-behavioural therapy (CBT) approach: OSVET is based on the CBT approach, which was chosen because it is one of the leading therapeutic approaches to dealing with distress overload and anxiety disorders. Previous studies show that it is the most suitable approach for algorithmic delivery style and for counselling with virtual assistants (Helgadottir 2018). Kessler et al. (2009) showed that online CBT is as effective as traditional 'in-person' therapy for the treatment of depression.
6. Assessment: OSVET first makes a screening of distress level based on the simple questionnaire. If the level is low, it provides some encouraging statements, directs the user towards simple relaxation techniques and reassures the user that his/her mental state is normal. If the level of distress is exceptionally high, the user is directed to get help with an in-person therapist and suggests an appropriate counsellor according to the detected problem. If there is a medium distress level, the virtual assistant continues with personalized conversation. There is also an option for the user to pass the initial screening and continue directly to a conversation with a virtual assistant.
7. Cognitive distortion identification: CBT typically focuses on a specific problem, helping the patient to identify, recognize and change disturbing

thought patterns and feelings that are leading to harmful or destructive beliefs and behaviours (Somers and Querée 2007). In OSVET, we identified the twelve mental distortions mentioned above. When a particular mental distortion is detected with the use of artificial intelligence algorithms, the user is asked to confirm the avatar's classification. If the user confirms it, (s)he is further guided to the appropriate personalized dialogue for the detected distortion. If the user does not confirm the way of thinking that avatar proposes, the avatar asks the user to describe in detail the mental problem and the loop repeats.

8. Personalized dialogue with adequate assignments: When the user's problem description is classified into one of the above cognitive distortions and confirmed by the user, one of the prepared scenarios, i.e., flexible sequences of dialogues is executed. It includes:

   - Psychoeducation: to educate users about the distorted way of thinking and representing the world is obviously of value.
   - Emotional support: i.e. supportive statements which encourage the users to deal with his/her problem and gives him/her hope to find a solution.
   - Guided self-help: OSVET guides users to think and analyse their way of thinking and mental representation. Several different therapeutic techniques are used, such as paraphrasing, active listening, etc.
   - Assignments: at the end of the conversation, assignments with instructions are presented to the user.
   - Redirection to a qualified in-person counsellor if necessary.

The data gathered throughout the conversation are anonymously saved in databases and regularly checked by the experts (see Fig. 5). If there are signs of misdiagnosis or further inappropriate dialogue, experts correct it, and thus OSVET is continuously improving.

The main benefits of the OSVET online counsellor are constant accessibility to all people in need, discretion and safety, professional, personalized treatment, based on natural language communication. As such, OSVET reduces stigma and encourages people in distress to search for help, leading to better mental health in a global society. There are also some disadvantages: the rapport is limited since the therapist cannot see the user, and thus cannot interpret facial cues, voice tone and body language. It is questionable if emotional depth achieved through written words can be as deep as in-person therapy. There is a particular risk of misdiagnosis. Even if safety is one of OSVETs main assumptions, there can still potentially be negative consequences when a system fails to understand emotions and incorrectly filters information during sensitive, high-stakes

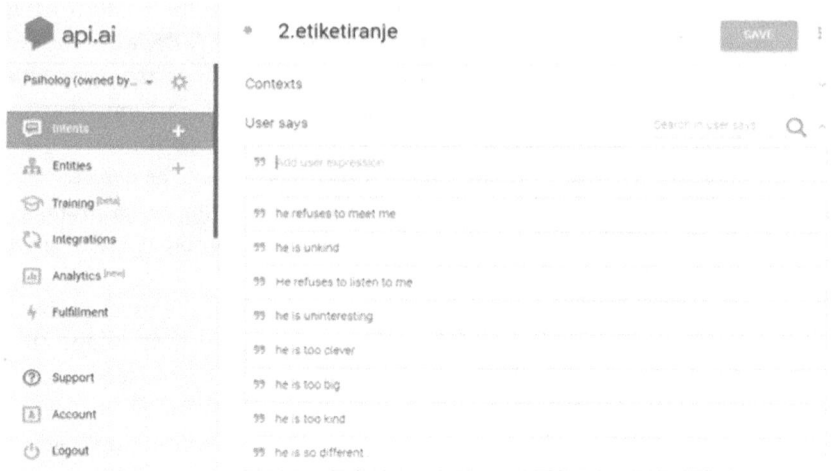

**Fig. 5:** OSVET database for the example of cognitive distortion 'global labelling'

conversations. As such, one should be aware of the fact that online counselling does not entirely substitute in-person therapy (Krivec & Gams 2017).

## Conclusion

We can conclude that Slovenian adolescents experience a great deal of stress, but they often do not know how to deal with stressful situations constructively. However, they express the need for learning the stress coping strategies that would enable them to experience less stress and better cope with it when needed. The usage of ICT for dealing with psychosocial problems is one of the possible strategies. Like all other, this strategy has its dis/advantages. To avoid possible disadvantages as much as possible, we propose the model of a personalized online counsellor (OSVET) to help adolescents dealing with their stressful situations and psychosocial problems.

OSVET was developed as an online psychosocial help solution for people in distress. The basic idea is derived from the fact that many people experience stress in their everyday life, but because of stigma they do not seek professional help. Moreover, technology has become an integral part of our lives. It seems inevitable that counselling will be affected by technology. Therefore, our goal was to develop the first stage of psychosocial help, which is accessible, professional and personalized and user-friendly. This innovation in psychosocial counselling is based on a combination of artificial intelligence and distance technology with elements of traditional psychotherapeutic techniques, e.g. expert knowledge stored in the form of scenarios. The approach could be beneficial in several fields, for students and pupils under distress, career counselling and similar. The OSVET system is, however, still in its initial developmental phase. We have: (1) analysed and conceptualized key concepts of the cognitive-behavioural counselling approach for the categorization of cognitive distortions from natural language, (2) formulated appropriate answers, exercises and instructions for the user, (3) implemented the content in an appropriate user interface, (4) tested the initial concept for half a year, corrected the basic mistakes and developed a basic knowledge base. The next step in OSVET development is the increase of its usage by different populations and widening the knowledge base on which the system works. An additional plan is also to link the OSVET solution to a network of therapists and psychosocial services providers to which the system will refer the user in need. Many other challenges include clinical, legal, technological, assessment and ethical issues. However, we believe that counselling solutions such as OSVET can be accepted as a useful clinical tool if only appropriate monitoring and connection is established with real therapists. In time, such solutions will encourage more people to start to resolve their mental health issues and consequently decrease stigma and promoting mental health hygiene in a global society.

## References

'7 Cups.' n.d. Accessed September 25, 2019. https://www.7cups.com/.

Addis, M. E., and J. R. Mahalik (2003). 'Men, Masculinity, and the Contexts of Help Seeking.' American Psychologist.

Andersson, G., P. Cuijpers, P. Carlbring, H. Riper, and E. Hedman (2014). 'Guided Internet-Based vs. Face-to-Face Cognitive Behavior Therapy for Psychiatric and Somatic Disorders: A Systematic Review and Meta-Analysis.' World Psychiatry : Official Journal of the World Psychiatric Association (WPA) 13 (3): 288–95.

Anye, E. T., T. L. Gallien, H. Bian, and M. Moulton (2013). 'The Relationship between Spiritual Well-Being and Health-Related Quality of Life in College Students.' Journal of American College Health : J of ACH 61 (7): 414–21.

Baghurst, T., and B. C. Kelley. (2014). 'An Examination of Stress in College Students Over the Course of a Semester.' Health Promotion Practice 15 (3): 438–47.

Beck, J. S. (2011). Cognitive Behavior Therapy: Basics and Beyond, 2nd Ed. Cognitive Behavior Therapy: Basics and beyond, 2nd Ed. New York, NY: Guilford Press.

'Best Counseling Degrees.' n.d. Accessed September 25, 2019. https://www.bestcounselingdegrees.net/online/.

'BetterHelp.' n.d. Accessed September 25, 2019. https://www.betterhelp.com/.

'Breakthrough: Confidential Online Counselling and Therapy.' n.d. Accessed September 25, 2019. https://www.breakthrough.com/.

'CBT Apps.' n.d. Accessed September 25, 2019. https://www.thriveport.com/.

Chang, T., C. J. Yeh, and J. D. Krumboltz (2001). 'Process and Outcome Evaluation of an On-Line Support Group for Asian American Male College Students.' Journal of Counselling Psychology 48 (3): 319–29.

Cohen, G. E., and B. A. Kerr (1999). 'Computer-Mediated Counselling.' Computers in Human Services 15 (4): 13–26.

Deal, S. L., and J. E. Williams (1988). 'Cognitive Distortions as Mediators between Life Stress and Depression in Adolescents.' Adolescence 23 (90): 477–90.

Dexter, L. R., K. Huff, M. Rudecki, and S. Abraham (2018). 'College Students' Stress Coping Behaviors and Perception of Stress-Effects Holistically.' International Journal of Studies in Nursing 3 (2): 1.

Dongier, M., R. Tempier, M. Lalinec-Michaud, and D. Meunier (1986). 'Telepsychiatry: Psychiatric Consultation through Two-Way Television. A Controlled Study.' Canadian Journal of Psychiatry. Revue Canadienne de Psychiatrie 31 (1): 32–34.

Fitzpatrick, K. K., A. Darcy, and M. Vierhile (2017). 'Delivering Cognitive Behavior Therapy to Young Adults with Symptoms of Depression and Anxiety Using a Fully Automated Conversational Agent (Woebot): A Randomized Controlled Trial.' JMIR Mental Health 4 (2): e19.

Glueckauf, R. L., S. P. Fritz, E. P. Ecklund-Johnson, H. J. Liss, P. Dages, and P. Carney (2002). 'Videoconferencing-Based Family Counseling for Rural Teenagers with Epilepsy: Phase 1 Findings.' Rehabilitation Psychology 47 (1): 49–72.

Helgadottir, F. (2018). 'Computerized Therapy: Will Your Next Therapist Be a Computer?' 2018. https://psychcentral.com/lib/computerized-therapy-will-your-next-therapist-be-a-computer/.

Hurst, C. S., L. E. Baranik, and F. Daniel (2013). 'College Student Stressors: A Review of the Qualitative Research.' Stress and Health : Journal of the International Society for the Investigation of Stress 29 (4): 275–85.

'ISHMO.' n.d. Accessed September 25, 2019. https://ismho.org/.

Jemec, J. (2019). 'Pozitivni Vplivi Informacijske in Komunikacijske Tehnologije Pri Soočanju Mladostnikov s Psihosocialnimi Težavami.' School of Advanced Social Studies.

Jeriček Klanšček, H., S. Roškar, K. Britovšek, N. Scagnetti, and M. Kuzmanič (2016). Mladostniki o Duševnem Zdravju. Ljubljana: Nacionalni inštitut za javno zdravje.

Jeriček Klanšček, H., H. Koprivnikar, A. Drev, V. Pucelj, T. Zupanič, and K. Britovšek (2015). Z Zdravjem Povezano Vedenje v Šolskem Obdobju Med Mladostniki v Sloveniji. Izsledki Mednarodne Raziskave HBSC, 2014. Ljubljana: Nacionalni inštitut za javno zdravje.

Kendall, P. C. (1992). 'Healthy Thinking.' Behavior Therapy 23 (1): 1–11.

Kessler, D., G. Lewis, S. Kaur, N. Wiles, M. King, S. Weich, D. J. Sharp, R. Araya, S. Hollinghurst, and T. J. Peters (2009). 'Therapist-Delivered Internet Psychotherapy for Depression in Primary Care: A Randomised Controlled Trial.' Lancet 374 (9690): 628–34.

Kowalski, R. M., G. W. Giumetti, A. N. Schroeder, and M. R. Lattanner (2014). 'Bullying in the Digital Age: A Critical Review and Meta-Analysis of Cyberbullying Research among Youth.' Psychological Bulletin 140 (4): 1073–1137.

Krivec, J., and M. Gams (2017). 'Virtual Personal Psychosocial Counsellor.' In Cognitonics: Proceedings of the 20th International Multiconference Information Society - IS 2017, 9th-13th October, edited by Vladimir FOMICHOV and Olga S. FOMICHOVA. Ljubljana.

Krivec, J., and P. Rakovec (2018). 'Cognitive Behavioral Profile of Stress Experience among Slovenian Students.' Kairos 12 (1–2): 67–86.

Krivec, J., and J. Suklan (2015). 'Vpliv Stigme Na Odnos Do Psihološke Pomoči.' Raziskave in Razprave 8 (3): 4–62.

Lazarus, R. S., and S. Folkman (1984). 'Stress, Appraisal, and Coping.' New York: Springer.

Lee, J., and S. Jang (2015). 'An Exploration of Stress and Satisfaction in College Students.' Services Marketing Quarterly 36 (3): 245–60.

Lekič, K., P. Tratnjek, N. Konec Jurčič, and M. Cugmas (2014). Srečanja Na Spletu. Potrebe Slovenske Mladine in Spletno Svetovanje. Celje: Nacionalni inštitut za javno zdravje.

Livingstone, S., L. Haddon, A. Görzig, and K. Ólafsson (2011). 'Risks and Safety on the Internet: The Perspective of European Children: Full Findings and Policy Implications from the EU Kids Online Survey of 9–16 Year Olds and Their Parents in 25 Countries.' London.

Love, B., B. Crook, C. M. Thompson, S. Zaitchik, J. Knapp, L. Lefebvre, B. Jones, E. Donovan-Kicken, E. Eargle, and R. Rechis (2012). 'Exploring Psychosocial Support Online: A Content Analysis of Messages in an Adolescent and Young Adult Cancer Community.' Cyberpsychology, Behavior, and Social Networking 15 (10): 555–59.

Lucas, G. M., J. Gratch, A. King, and L. P. Morency (2014). 'It's Only a Computer: Virtual Humans Increase Willingness to Disclose.' Computers in Human Behavior 37: 94–100.

Machin, M. A., and P. Creed (1999). 'Changing Wonky Beliefs Training Program.' University of Southern Queensland.

Makarovič, M., Macur, M., and Rončević, B. (2011). 'Policy Challenges of Problem Gambling in Slovenia'. Ljetopis socijalnog rada, 18 (1): 127–152.

Mallen, M. J., D. L. Vogel, A. B. Rochlen, and S. Day (2005). 'Online Counseling: Reviewing the Literature from a Counselling Psychology Framework.' The Counselling Psychologist 33 (6): 819–71.

'Metanoia: Online Therapy.' n.d. Accessed September 25, 2019. https://www.metanoia.org/.

'PARO Therapeutic Robot.' n.d. Accessed September 25, 2019. http://www.pararobots.com/.

Piatetsky-Shapiro, G. (1991). Discovery, Analysis, and Presentation of Strong Rules. In G. Piatetsky-Shapiro, and W. J. Frawley (ed.), Knowledge Discovery in Databases, AAAI/MIT Press, Cambridge, MA.

Powell, J., S. Martin, P. Sutcliffe, D. Todkill, E. Gilbert, M. Paul, and J. Sturt (2010). 'Young People & Mental Health: The Role of Information and Communication Technology.'

Prensky, M. (2012). From Digital Natives to Digital Wisdom: Hopeful Essays on Education. Corwin Press.

Rončević, B., and Tomšič, M. (2017). 'Perspectives of Information Society: Bricolage of Manifestations'. In: Rončević, B. and Tomšič, M. (eds.). Information Society and Its Manifestations: Economy, Politics, Culture. Frankfurt am Main: Peter Lang, doi: 10.3726/b10694.

Rus-Calafell, M., J. Gutiérrez-Maldonado, C. Botella, and R. M. Baños (2013). 'Virtual Reality Exposure and Imaginal Exposure in the Treatment of Fear of Flying: A Pilot Study.' Behavior Modification 37 (4): 568–90.

'Samaritans.' n.d. Accessed September 25, 2019. https://www.samaritans.org/.

Senson, A. (2016). 'Virtual Reality Therapy: Treating The Global Mental Health Crisis | TechCrunch.' https://techcrunch.com/2016/01/06/virtual-reality-therapy-treating-the-global-mental-health-crisis/?guccounter=1&guce_

referrer_us=aHR0cHM6Ly93d3cuZ29vZ2xlLmNvbS88&g
uce_referrer_cs=8z7RV-7ud4tgwjxXAfqPbw.

Somers, J., and M. Querée (2007). Cognitive Behavioural Therapy. Vancouver, B.C: Centre for Applied Research in Mental Health and Addictions (CARMHA).

Suler, J. R. (2000). 'Psychotherapy in Cyberspace: A 5-Dimensional Model of Online and Computer-Mediated Psychotherapy.' Cyberpsychology and Behavior 3 (2): 151–59.

'Talkspace.' n.d. Accessed September 25, 2019. https://www.talkspace.com/.

U.S. Department of Health and Human Services (1999). 'Mental Health: A Report of the Surgeon General.' Rockville, MD.

Wagner, B., A. B. Horn, and A. Maercker (2014). 'Internet-Based versus Face-to-Face Cognitive-Behavioral Intervention for Depression: A Randomized Controlled Non-Inferiority Trial.' Journal of Affective Disorders 152–154 (January): 113–21.

'Woebot.' n.d. Accessed September 25, 2019. https://woebot.io/.

# Part 2 Challenges of Learning and Entrepreneurship

Petra Kleindienst and Andrej Raspor

# The Effectiveness of E-learning for Students' Acquiring Online Presentation Competence

**Abstract:** Presenting information effectively is a vital element in the crucial moments of one's professional life. This is especially relevant for students who are laying the groundwork for their presentation competence while they are studying. Since information and communication technologies (ICT) have become popular among students and because learning material and lectures are often delivered to students using online tools, it is essential to explore the development of presentation competence in a virtual learning environment. This chapter focuses on the research question about the (dis)advantages of e-learning environments in terms of building students' online presentation competence. This research is based on a case study of a higher education institution in Slovenia. It measures the improvement of students' competence during their assignment presentations in an online study course. The purpose of this chapter is to present the results of students' self-evaluations of their presentation competence before the start and after the completion of the online course. The research results show that students have considerably improved their presentation competence by conducting online presentations.

**Keywords:** competences, presentations, study, e-learning, experiential learning, self-assessment

## 1. Introduction

Presentations are one of the modes for providing experiential learning opportunity (Ord 2009). As delivering valuable presentations can be regarded as one of the most significant challenges for students, this challenge is even more daunting when presentations are conducted using online tools, especially when using a camera. Such presentations are demanding for both pedagogues and students. Pedagogues must adjust their evaluating system and instructions to online presentations so that they are comparable with face-to-face presentations. According to Mitchell-Holder (2016), communicating with students in an online environment requires more thought and planning than doing so in a face-to-face environment due to the absence of body language in the online environment. In contrast, students are supposed to familiarize themselves with the critical elements of online presentations and prove their abilities to have effective online oral communication when using online presentation tools. Online presentations differ from face-to-face presentations: students have to acquire additional skills and competences to make their presentations effective. There are diverse types of online presentations, including slide decks, audio presentations, video presentations. For the purpose of this chapter, online presentations shall be considered interactive video presentations performed using an online collaborative tool; providing audio presence and allowing interactions with a live audience as well as using chat options and PowerPoint presentations. Al Tawil (2009) suggests that online learning is still regarded as less satisfactory than face-to-face learning environments. The reason for this is not the decreased level of education, but rather the decreased level of human interaction. However, according to Hampel (2013), online communication may be liberating and confidence-building in some cases. When engaging in online communication, students feel less pressure and are more prepared to take greater risks and lower their affective filter. Consequently, online environments can have a positive impact on the learner's identity, self-image and motivation (Dörnyei 2005; Hampel 2013).

Students frequently face difficulties when adapting to online presentations. However, the presentations can bring valuable results, especially in relation to building and enriching students' presentation competence. Although in recent decades a significant among of published research has been devoted to the challenges of e-learning online learning tools and delivering knowledge in the digital age (for example, Andrews and Haythornthwaite 2016; Hamad 2017; Khan and Ally 2015; Krivec 2017; Pelet 2019; Rosenberg 2001), few studies have examined the competences acquired during online presentations. Therefore, we focus on the research question: What are the (dis)advantages of e-learning environment in terms of building students' online presentation competence?

To answer that question, a case study of a higher institution in Slovenia that implements online students' presentations into the learning environment was conducted. The study is based on the quantitative methodology (i.e., the questionnaires filled by the students who have experienced an online study course and online presentations). The study aimed to detect (dis)advantages of implementing online presentations in the learning environment. More specifically, it aimed to identify the competences that are essentially improved or hindered after experiencing online presentations. The case study was conducted in an intercultural environment, at a Slovenian higher education institution that attracts students from three markets (Slovenia, Croatia and Serbia). The survey was conducted in the national languages of the students. Students enrolled in the study programme as well as those who have already completed the programme (not more than two years after the graduation) participated in the survey.

Digital information and communication technologies are the primary engines of the information society in various areas (Rončević and Tomšič 2017;

Rončević et al. 2019; Fric et al. 2020; Fric and Rončević 2018; Mileva Boshkoska et al. 2018); globalization, in conjunction with technological advancement, has significantly influenced the character of societies (Adam and Tomšič 2019) as well as the educational system (Golob 2017; Golob and Makarovič 2018; Golob and Makarovič 2019) or innovation performance (Modic and Rončević 2018; Rončević and Modic 2011). Research and practice suggest different definitions and classifications of digital skills and competence (DSC). An emerging classification in the EU identifies three main categories of DSC for learners/citizens: digital competence, job-specific digital skills and digital skills for ICT professionals. Digital competence is often referred to as 'digital literacy' (for a discussion on the meaning of those two terms, see Spante et al. 2018). Digital competence encompasses a set of necessary digital skills, covering information and data literacy, online communication and collaboration, digital content creation, safety and problem solving. Digital competence is about the ability to apply those digital skills (knowledge and attitude) in a confident, critical and responsible manner in a defined context (e.g. education). Since 2006, digital competence has been one of the eight key competences in the EU for lifelong learning (Brolpito 2018).

## 2. Theoretical framework

The ability to make a presentation is undoubtedly one of the essential competencies for professionals and their effective performance at the workplace, their successful interaction and business conduct (Bourhis and Allen 1998; De Grez 2009; Kerby and Romine 2009; Van Ginkel et al. 2015). However, presenting by individuals is still related with fear in social contexts (Smith and Sodano 2011); a potential means to reduce fear when presenting is to develop students' oral presentation competence (Van Ginkel et al. 2019). For this chapter, we refer to De Grez' definition of oral presentation competence as 'a combination of knowledge, skills and attitudes needed to speak in public in order to inform, self-express, relate, or to persuade' (De Grez 2009, 5). Oral presentations allow direct interaction with the audience. However, one must be able to deliver the presentation effectively to make it compelling and informative. An effective presentation can help secure a job, seal a deal with a client and more (Greenlaw 2012).

The Constructivist theory implies that active learning (i.e., learning that stimulates learners to play active roles) encourages students to learn more effectively than passive learning does. Individuals are assumed to learn better when they do research by themselves and when they control the pace of learning (Leidner and Jarvenpaa 1995). Students conducting presentations indicates active learning and presumably leads to better presentation competence. Presentations positively stimulate academic debates while motivating students' active learning through preparation, presentation, defence, as well as classmates' participation and interaction. Those in the debate audience also learn a great deal from their observations (Snider and Maxwell 2002). Engagement is related to motivation and student achievement. Greater engagement is associated with higher levels of academic achievement (Chafouleas 2019).

Due to the logistics involved, oral presentations are often eliminated from online courses (Kenkel 2011). One cannot disregard the fact that e-learning technologies have powerful influences on the transformations of the educational system. Education has moved into a new era of profound change related to the designing and delivering of e-learning programmes (Kanuka 2008). Popular e-learning management systems include Moodle, Blackboard, Desire2Learn, among others. Many new possibilities in education have become apparent, as well as many new challenges (ibid.).

One critique of students using online presentations is that a significant part of effective communication is transmitted through body language and other forms of non-verbal behaviour (Mehrabian 1972; Mehrabian and Ferris 1967; Mehrabian and Wiener 1967). Non-verbal behaviour has long been identified as a significant phenomenon (Argyle 1969; Birdwhistell 1970; Harisson 1973; Piaget 1960; Poyotos 1977; Harper, Wiens and Matarazzo 1978) and is still receiving significant attention at the academic level. Non-verbal behaviour can be classified as the verbal-vocal, nonverbal-vocal and the nonverbal-non-vocal (Mandal 2014). According to Mahrabian (1972; see also Mehrabian and Wiener 1967; Mehrabian and Ferris 1967), 7 % of a message pertains to feelings and attitudes is in the words that are spoken; 38 % pertains to the feelings and attitudes and is paralinguistic (way that the words are said); 55 % pertains to feelings and attitudes in facial expression.

Body language and facial expression are excluded from e-study environments, which is often described as one of the disadvantages in e-learning environments; one could say that e-learning environment lacks a richness of communication (Vrasidas and Zembylas 2003; Moon, Birchall, Williams and Vrasidas 2005). However, despite the lack of the impacts of the presenter's eye contact, facial expressions, hand gestures and body movements, a student can still operate with speech rhythm, volume and voice intonation in an online presentation. Additionally, the focus on the quality of presentations from the content point of view is more intensive in e-environment, as the presenter cannot rely on body movements and gestures. However, many academic institutions are developing online degree programmes. Thus, e-learning is demanding that students learn new approaches to present their assignments; in other words, they must acquire new competences to conduct online presentations. Furthermore, students are continually gaining additional new competences when preparing for online presentations, which means a new stage of acquiring presentation competence during their studies.

Following from the findings stated above, having online presentations instead of face-to-face presentations certainly has some noticeable disadvantages. However, the technology is currently developing at a rapid pace; it has changed the world drastically in the last two decades. In light of the rate of technological development, student presentations are becoming a common practice in many online environments. However, online presentations are often not well accepted among students. Therefore, there is a need that students recognize the advantages of having the opportunity to conduct online presentations and absorb new presentation competence that can be gained in e-learning environments.

As the students of different nationalities (Slovenians, Croatians and Serbians) were included in the study, it is relevant to emphasize that one can perceive the cultural-specific differences in body language. According to Busà (2015), for example, the use of frequent, broad, full-arm, animated gesturing during speech may be common in some communities. In others, it may be considered distracting, may annoy the listener and ultimately affect the image the speaker is projecting of him/herself (ibid.). Slovenia, Croatia and Serbia, as the former republics of Yugoslavia, have some common characteristics (Iglar 1992): however, there is a lack of scientifically supported research on differences in body language between those three countries. In this regard, opportunities for future research remain.

## 3. Research Design

The study was carried out at a higher institution in Slovenia (DOBA), which runs an online study course. Students enrolled in the online course belong to the different nationalities (Slovenian, Croatian and Serbian). The survey was conducted in the national languages of the students. We consider a selected case study as relevant for our research as students from those three countries are frequently not keen to do online presentations. In general, they still consider face-to-face presentations to be more effective than online presentations (Campbell 2015). Students who are not confident enough to speak in front of others may not feel comfortable participating in online discussions (Thor et al. 2017). They are reticent to presenting their assignments in general; one can often perceive a lack of interest and self-confidence in students when conducting presentations online. Presenting in front of an audience, regardless of the mode (online or face-to-face), can create anxiety (Campbell 2015). When involved in the e-learning environment, students are often especially reserved about using the camera during their presentations. Audio online presentations using no camera are usually easier to perform by students. Our research question is: What are the (dis)advantages of e-learning environments in terms of building students' online presentation competence?

The part-time students of an online master study programme (enrolled 2013–2017) who have experienced online presentations via Blackboard Collaborate were included in the study. Blackboard Collaborate is a virtual online classroom tool in which students and pedagogues interact synchronously with one another; in line with the existing studies, it can bring many valuable advantages to educational settings (Hart, Bird and Farmer 2019). The pedagogue leading the session can upload pre-prepared content, for example, PowerPoint presentations, documents and similar, stream live video content from a webcam and share their screen with participants (e.g., for software training) as well as share the audio by his/her microphone. In addition, the pedagogues can make use of a digital whiteboard, run polls and quizzes and chat with participants. The students can also share their audio and webcam feeds during the session, as well as presenting and sharing their screens during the session (ibid.).

The research is based on quantitative methods. It was conducted in 2016–2017, using the 1ka.com website. All the participants of the study were asked to complete the questionnaire measuring students' presentation competence before and after the completion of the online course. Thus, the results about the acquired students' presentation competence are based on self-assessments. More specifically, the survey was aimed to measure the students' self-perceived competences before the start and after the completion of the online course. Following that, the improvement of students' presentation competence during the active involvement in the online presentations was measured. Fourteen statements were designed to measure the students' behaviour by using quantitative methods. A 7-point Likert scale was used, ranging from 1 ('no experience') to 7 ('well experienced'):

1. I create good PowerPoint presentations (concise, comprehensive, without too much text, visually attractive, with animations).
2. I use key elements of a good presentation (sound and clear speech, self-confidence, concise expression, no stuttering or buzzwords, appropriate body language).

3. I easily use the Blackboard Collaborate tool to present tasks, projects and ideas.
4. I use the camera in presentations via Blackboard Collaborate.
5. I finish the presentation on time.
6. I encourage discussion after my presentation.
7. I am confident when presenting via Blackboard Collaborate.
8. Blackboard Collaborate enables me to make an effective presentation of my idea and successfully communicate with my colleagues and professor.
9. I identified possible questions from the professor and colleagues and prepared answers.
10. I checked all the equipment before the presentation, and it was functional.
11. I receive negative feedback as encouragement for improvement.
12. I enrol in the discussions.
13. I provide my feedback with the sole intention of helping others to improve.
14. I provide positive feedback and encouragement to my colleagues.
15. I can evaluate the presentations of my study colleagues.

The questions were designed according to the required online skills to prepare a presentation, conduct a presentation and encourage participants.

One hundred and thirty-five valid students' answers were gathered during the research; missing data were ignored and not included in the research calculations (Tab. 1). The majority of respondents were female (60 %) and between twenty-five and forty years old (57.4 %); 37.5 % of respondents were older than forty years. Almost all participants were employed (93 %). Half of them were enrolled in the 2015 academic year, while the rest of them were enrolled in the 2013, 2014 and 2016 academic years.

## Research results

Table 2 shows the survey results referring to the findings regarding how students have evaluated the increase in individual competencies before the start and after the completion of the online course.

On average, students' rate improvement 'before-after' at 19 %. The most improved behaviour before the start and after the completion of the online course is 'I provide positive feedback and encouragement to my colleagues' (34 %), followed by 'I easily use the Blackboard Collaborate tool to present tasks, projects and ideas' (30 %) and 'I use the camera in presentations via Blackboard Collaborate' (27 %). According to the survey, the significant opportunities for improvement are connected with the following statements: 'I provide my feedback with the sole intention of helping others to improve' (improvement before-after

**Tab. 1:** Demographic information

| Gender | Slovenia | Other markets | Total | Valid per cent |
|--------|----------|---------------|-------|----------------|
| Male | 29 | 25 | 54 | 40 % |
| Female | 56 | 25 | 81 | 60 % |
| Valid | 85 | 50 | 135 | 100 % |
| **Age** | **Slovenia** | **Other markets** | **Total** | **Valid per cent** |
| 1 (24 years) | 4 | 3 | 7 | 5.1 % |
| 2 (25–40 years) | 44 | 34 | 78 | 57.4 % |
| 3 (40 or older) | 37 | 14 | 51 | 37.5 % |
| Valid | 85 | 51 | 136 | 100 % |
| **The current status** | **Slovenia** | **Other markets** | **Total** | **Valid per cent** |
| employed | 78 | 48 | 126 | 93 % |
| unemployed | 7 | 3 | 10 | 7 % |
| retired | 0 | 0 | 0 | 0 % |
| Valid | 85 | 51 | 136 | 100 % |
| **Year subscription** | **Slovenia** | **Other markets** | **Total** | **Valid per cent** |
| enrolled in the 2013 academic year | 19 | 4 | 23 | 17 % |
| enrolled in the 2014 academic year | 24 | 11 | 35 | 26 % |
| enrolled in the 2015 academic year | 41 | 27 | 68 | 50 % |
| enrolled in the 2016 academic year | 0 | 9 | 9 | 7 % |
| Valid | 84 | 51 | 135 | 100 % |

only 9 %); 'I received negative feedback as encouragement for improvement' (improvement before-after only 11 %); 'I checked all technical equipment before the presentation, and it was functional' (improvement before-after only 14 %).

Pairwise t-tests were run to check for differences between before and after the implementation of online lectures (Tab. 3). All pairwise t-tests showed a statistically significant difference ($\alpha$=0.05) between before and after the lectures. There are statistically significant differences across all parameters. However, one should be careful generalizing from research findings as the research data were not collected at the level of the entire country but rather at the level of one higher education institution.

## Interpretation

Sometimes a lack of relevant technical skills of learners can hinder the success of online study course (Pathak and Vyas 2019). However, the present research showed a definite improvement in the ability of students to use the Blackboard

**Tab. 2:** Improving individual competencies

| BEHAVIOUR | IMPROVEMENT (BEFORE-AFTER) |
|---|---|
| I create good PowerPoint presentations (concise, comprehensive, without too much text, visually attractive, with animations). | 15 % |
| I use key elements of a good presentation (sound and clear speech, self-confidence, concise expression, no stuttering or buzzwords, appropriate body language). | 14 % |
| I easily use the Blackboard Collaborate tool to present tasks, projects and idea. | 30 % |
| I use the camera in presentations via Blackboard Collaborate. | 27 % |
| I finish the presentation on time. | 15 % |
| I encourage discussion after my presentation. | 17 % |
| I am confident when presenting via Blackboard Collaborate. | 30 % |
| Blackboard Collaborate enables me to make an effective presentation of my idea and successfully communicate with my colleagues and professor. | 25 % |
| I identified possible questions from the professor and colleagues and prepared answers. | 21 % |
| I checked all the equipment before the presentation, and it was functional. | 14 % |
| I received negative feedback as encouragement for improvement. | 11 % |
| I enrolled in the discussions. | 21 % |
| I provide my feedback with the sole intention of helping others to improve. | 9 % |
| I provide positive feedback and encouragement to my colleagues. | 34 % |
| I can evaluate the presentations of my study colleagues | 15 % |
| **Average** | **19 %** |

tool (improvement 30 %). That was an expected result. After the completion of the online course, students assessed that they more easily use an online tool (Blackboard Collaborate) to present projects, tasks and ideas.

Online presentations encourage students to use the camera more often to represent their assignments (improvement 27 %). When enrolled in non-online study programmes, students are used to making presentations in front of others; they consider the conduct of the presentation to be more or less personal. When being enrolled in e-learning, students tend to reach a satisfying level of contact with the other participants, which is not possible when the camera is off. Thus, the camera fosters a presentation and especially a discussion among students

**Tab. 3:** T-test

| | | Difference between the means | Std. Deviation | N | p |
|---|---|---|---|---|---|
| Pair 1 | I create good PowerPoint presentations (concise, comprehensive, without too much text, visually attractive, with animations). | 0.83 | 1.32 | 123 | < 0.001 |
| Pair 2 | I use key elements of a good presentation (sound and clear speech, self-confidence, concise expression, no stuttering or buzzwords, body language). | 0.85 | 1.45 | 123 | < 0.001 |
| Pair 3 | I easily use the Blackboard Collaborate tool to present tasks, projects and ideas. | 1.46 | 1.89 | 122 | < 0.001 |
| Pair 4 | I use the camera in presentations via Blackboard Collaborate. | 1.02 | 1.89 | 122 | < 0.001 |
| Pair 5 | I finish the presentation in time. | 0.71 | 1.39 | 121 | < 0.001 |
| Pair 6 | I encourage discussion after my presentation. | 0.71 | 1.15 | 121 | < 0.001 |
| Pair 7 | I am confident when presenting via Blackboard Collaborate. | 1.25 | 1.58 | 122 | < 0.001 |
| Pair 8 | Blackboard Collaborate enables me to make an effective presentation of my idea and successfully communicate with my colleagues and professor. | 1.25 | 1.81 | 122 | < 0.001 |
| Pair 9 | I identified possible questions from the professor and colleagues and prepared answers. | 0.92 | 1.34 | 122 | < 0.001 |
| Pair 10 | I checked all equipment before the presentation, and it was functional. | 0.79 | 1.49 | 122 | < 0.001 |
| Pair 11 | I received negative feedback as encouragement for improvement. | 0.54 | 1.13 | 121 | < 0.001 |

**Tab. 3:** Continued

| | Difference between the means | Std. Deviation | N | p |
|---|---|---|---|---|
| Pair 12 I enrolled in the discussions. | 0.97 | 1.52 | 122 | < 0.001 |
| Pair 13 I provide my feedback with the sole intention of helping others to improve. | 0.50 | 1.19 | 121 | < 0.001 |
| Pair 14 I provide positive feedback and encouragement to my colleagues. | 0.59 | 1.14 | 122 | < 0.001 |
| Pair 15 I can evaluate the presentations of my study colleagues | 0.81 | 1.09 | 73 | < 0.001 |

and educator to be more personal, enabling a presenter being in touch with the others who are present at a lecture. Additionally, a camera provides authenticity to student's work.

Furthermore, the significant improvement before and after the use of an online tool (34 %) is a result of students providing positive feedback and encouragement to their colleagues. This result implies that students are able to provide positive feedback and encouragement to their colleagues more effectively than before using online presentations. This means that online programmes encourage students to foster their engagement in a discussion. When students do not pay a great deal of attention to the others' appearance and their non-verbal behaviour or body language, their focus is more on the content of the presentation and the discussion. Thus, online presentations encourage students to give more constructive remarks, questions, comments and similar. This is in line with the findings of Pathak and Vyas (2019) who stress the advantages of students' ability to interact with others and resolve their doubts, if any, when being enrolled in online courses. However, an e-learning environment often contributes to decreasing face-to-face social engagement as students might not socialize with other students as in a face-to-face learning environment.

Additionally, students often miss the physical connection between students and faculty members in such an environment (Holley and Taylor 2009). The research shows that this does not diminish the importance of active, collaborative learning for online curricula (ibid.). According to Palloff and Pratt (2005), key to the learning process in e-learning environments is interactions among students themselves, interactions between students and faculty, and the collaboration in learning that results from their interaction.

According to Pathak and Vyas (2019), a disadvantage of online presentations as well as online study courses, as such, is that students are obtaining merely theoretical knowledge. At the same time, students might not receive sufficient knowledge to use what they have theoretically learnt during their online course practically; face-to-face learning experience might be missing when enrolled in an on-line course (ibid.). However, as the content of the presentation is much more in the foreground in the online presentations due to the limited meaning of body language, a student is supposed to make a more significant effort from the content point of view when preparing online presentations than face-to-face presentations, in order to provide a concise and understandable presentation. In online presentations, elements such as the tone of the voice, are considerably more critical in order to ensure an effective presentation. Thus, we can conclude that online presentations enrich students with new competencies in terms of providing presentations that are powerful in the view of the content but conducted with less use (help) of body language than face-to-face presentations. The main focus of an online presentation is, therefore, primarily on the content.

Following from that, online presentations boost academic debates and stimulate participants to be 'active students'. This is also because students do not focus on visual images through online presentations as much as through face-to-face presentations. When they do not pay attention to the appearance of others, they focus more on the content of the presentation and the discussion, so they can give more constructive comments. This is relevant as the feedback is one of the mirrors of students' growth (Pathak and Vyas 2019). Additionally, the research results show that students' ability to identify possible questions from the professor and colleagues and prepare answers in advance improves when having online presentations (by 21 %). However, a threat that students might not have enough time to work with gained feedback properly still exist, which could lead to some students falling behind, having gaps in their knowledge and not completing the course successfully (Pathak and Vyas 2019).

However, opportunities for improvements remain. For example, the research indicates a relatively low level of improvement (by 11 %) of the ability to positively accept critique before and after using an online tool. Students are still not eager to receive negative feedback, which could be related to their emotional processing skills (Stepišnik Perdih 2018). Furthermore, despite the fact that students assess their abilities to make presentations better than before using the online tool (sound and clear speech, self-confidence, concise expression, no stuttering or buzzwords, body language), there are still many opportunities to improve as well (improvement only 14 %).

Additionally, there are opportunities to persuade students that providing feedback to the others is intended to help others (improvement only 9 %). The latter result implies the personal values of students included in our survey, namely a relatively low level of benevolence values. An intention to help others is related to benevolence or self-transcendence personal values (Sagiv et al. 2017). The importance of benevolence (self-transcendence) values correlates positively with the likelihood of engaging in helpful acts and can bring more altruistic behaviour (Sagiv et al. 2017). Therefore, we can conclude that upgrading the benevolence personal values of students could bring improvement of students' behaviour measured by our survey.

## Conclusion

E-Learning is a very popular and significant subset in education technology, because it offers an online learning and teaching platform to specific knowledge through the help of internet technology in all over the world (Singh et al. 2017). Our research showed that e-learning and online presentations make students pay more attention to the content and less on external elements (for example, body language, outer appearance, dress, gestures, etc.). From this, we can conclude that e-learning stimulates students to become more precise, cautious and attentive to the content of the presentation, thus enriching students with new competences. Additionally, our research reveals that online presentations in e-learning study environment improves at least a part of (online communication and collaboration, digital content creation) digital and online learning competence recommended by the EU (Brolpito 2018). It must be emphasized that the research results show improvements in all the measured indicators. The research results undoubtedly show that online presentations enhance the students' engagement in the discussion as well as encouraging students to become more active.

The technology might explain the observed differences in online courses, which permits students to correct mistakes and re-record a presentation before submitting it, or the larger withdrawal rate which may selectively remove those students who may have done poorly in either format (Broeckelman-Post et al. 2019).

However, the disadvantages of e-learning and online presentation are, inter alia, that students who are not willing to attend the discussion have more chances to be passive, either because they are reserved or because they are not interested in the topic or lectures as such. Particularly important here is the role of the moderator in ensuring the activity of all students.

The recommendation for online education providers is to give attention to how students deliver presentations and give them feedback. Due to the weakness of our research, which is based on self-assessment, the recommendation for future researchers is to survey random students before the first encounter with an online presentation (the first subject) and after they have already become experienced users (presentation of the final thesis), combining self-assessment and professors' assessment method. This could upgrade our research results, offering a more accurate picture of how much students have improved their competencies during on-line study.

## Bibliography

Adam, Frane, Tomšič, Matevž, 'The Future of Populism in a Comparative European and Global Context', Comparative sociology 18/5–6 (2019), 687–705.

Al Tawil, Rima, 'Nonverbal Communication in Text-Based, Asynchronous Online Education', International Review of Research in Open and Distributed Learning 20/1 (2019). <http://www.irrodl.org/index.php/irrodl/article/view/3705/4961>, accessed 1 September, 2019.

Andrews, Richard and Caroline Haythornthwaite, eds, The SAGE Handbook of E-learning Research, 1st Edition (Los Angeles/London/New Delhi/Singapore/Washington DC/Melbourne: SAGE Publications Ltd, 2016).

Argyle, Michael, Social Interaction (New York, NY: Atherton, 1969).

Birdwhistell, Ray L., Kinesics and Context. Essays on Body Motion Communication (Philadelphia: University of Pennsylvania Press, 1970).

Bourhis, John, and Mike Allen. 1998. 'The Role of Videotaped Feedback in the Instruction of Public Speaking: A Quantitative Synthesis of Published Empirical Research', Communication Research Reports 15/3 (1998), 256–61, Doi:10.1080/08824099809362121.

Broeckelman-Post, Melissa A.; Katherine E. Hyatt Hawkins, Anthony R. Arciero, and Andie S. Malterud, Online versus Face-to-Face Public Speaking Outcomes: A Comprehensive Assessment, Basic Communication Course Annual 31/1 (2019), 10.

Brolpito, Allesandro, Digital Skills and Competence, and Digital and Online Learning (Turin: European Training Foundation, 2018).

Busà, Maria Grazia, 'Teaching learners to communicate effectively in the L2, Integrating body language in the students' syllabus', Lingue e Linguaggi 15 (2015), 83–98.

Campbell, Scott, 'Presentation Anxiety Analysis: Comparing Face-to-Face Presentations and Webinars', Journal of Case Studies in Education 7 (2015).

Chafouleas, Sandra M., Taylor Koriakin, Katrina D. Roundfield, Stacy Overstreet, 'Addressing Childhood Trauma in School Settings: A Framework for Evidence-Based Practice', School Mental Health 11/1 (2019), 40–53.

De Grez, Luc, 'Optimizing the Instructional Environment to Learn Presentation Skills', PhD thesis (University of Gent, 2009).

Dörnyei, Zoltán, The Psychology of the Language Learner: Individual Differences in Second Language Acquisition (Mahwah, NJ: Lawrence Erlbaum, 2005).

Fric, Urška, Rončević, Borut, 'E-simbioza: Leading the Way to a Circular Economy through Industrial Symbiosis in Slovenia', Socijalna ekologija: časopis za ekološku misao i sociologijska istraživanja okoline, 27/2 (2018), 119–140, doi: 10.17234/SocEkol.27.2.1.

Fric, Urška, Rončević, Borut, Džajić Uršič, Erika, 'Role of Computer Software Tools in Industrial Symbiotic Networks and the Examination of Socio-Cultural Factors', Environmental Progress & Sustainable Energy, 39 (2020), e13364, doi: 10.1002/ ep.13364.

Greenlaw, Raymond, Technical Writing, Presentational Skills, and Online Communication: Professional Tools and Insights (IGI Global, 2012).

Golob, Tea, 'Evropska študijska mobilnost kot sodobni obred prehoda', Glasnik Slovenskega etnološkega društva 57/3–4 (2017), 75–84.

Golob, Tea and Makarovič, Matej, 'Student Mobility and Transnational Social Ties as Factors of Reflexivity', Social sciences 7/3 (2018), 1–18.

Golob, Tea and Makarovič, Matej, 'Reflexivity and Structural Positions: The Effects of Generation, Gender and Education', Social sciences 8/9 (2019), 1–23.

Hamad, Mona M. 'Pros & Cons of Using Blackboard Collaborate for Blended Learning on Students Learning Outcomes', Higher Education Studies 7/2 (2017), 7–16.

Hampel, Regine, 'Making Meaning Online: Computer-Mediated Communication for Language Learning', in Peti-Stantic, Anita and Stanojevic, Mateusz-Milan, eds, Language as Information. Proceedings from the CALS Conference 2012 (Frankfurt am Main: Peter Lang, 2013).

Harper, Robert G., Arthur N. Wiens, and Joseph D. Matarazzo, Nonverbal communication: The state of the art (New York, NY: John Wiley & Sons, 1978).

Harrison, Randall P., Nonverbal communication, in I. de Solo Pool, W. Schramm, N. Maccoby, F. Fry, E. Parker, and J. L. Fein, eds, Handbook of communication, 46–76 (Chicago, IL: Rand McNally, 1973).

Hart, Tania, David Bird and Robert Farmer, 'Using Blackboard Collaborate, a Digital Web Conference Tool, to Support Nursing Students Placement Learning: A Pilot Study Exploring Its Impact', Nurse Education in Practice 38 (2019), 72–78.

Holley, Karri A. and Taylor, Barret J., 'Undergraduate Student Socialization and Learning in an Online Professional Curriculum', Innov High Educ 33 (2009), 257–269.

Iglar, Richard F., 'The Constitutional Crisis in Yugoslavia and the International Law of Self-Determination: Slovenia's and Croatia's Right to Secede', Boston College International and Comparative Law Review 15/1 (1992).

Kanuka, Heather, 'Understanding e-Learning Technologies-in-Practice through Philosophies-in-Practise', in Terry Anderson, ed., The Theory and Practice of Online Learning (Edmonton: AU Press, 2008).

Kenkel, Cindy S., 'Teaching Presentation Skills in Online Business Communication', Journal of Online Learning and Teaching 7/3 (2011), 412–418.

Kerby, Debra, and Jeff Romine, 'Develop Oral Presentation Skills Through Accounting Curriculum Design and Course-Embedded Assessment', Journal of Education for Business 85/3 (2009), 172–79, Doi:10.1080/08832320903252389.

Khan, Badrul H. and Mohamed Ally, International Handbook of E-Learning, Volume 1: Theoretical Perspectives and Research (New York and London: Routledge, 2015).

Krivec, Jana, 'Cognitive processes and information technology in education', in Rončević, Borut and Matevž Tomšič, eds, Information society and its manifestations: economy, politics, culture (Frankfurt am Main [etc.]: PL Academic Research, 2017).

Leidner, Dorothy E., and Jarvenpaa, Sirkka L. 1995. 'The Use of Information Technology to Enhance Management School Education: A Theoretical View', MIS Quarterly: Management Information Systems 19/3 (1995), 265–291.

Mandal, Fatik Baran, 'Nonverbal Communication in Humans', Journal of Human Behavior in the Social Environment 24/4 (2014), 417–421, Doi: 10.1080/10911359.2013.831288.

Mehrabian, Albert, Non-verbal communication (Chicago, Illinois: Aldine- Atherton, 1972).

Mehrabian A. and Ferris S. R. 'Inference of Attitudes from Nonverbal Communication in Two Channels', Journal of Consulting Psychology 31/3 (1967), 48–258.

Mehrabian A. and Wiener M., 'Decoding of Inconsistent Communications', Journal of Personality and Social Psychology 6 (1967), 109–114.

Mileva Boshkoska, Biljana, Rončević, Borut, and Džajić Uršič, Erika, 'Modeling and Evaluation of the Possibilities of Forming a Regional Industrial Symbiosis Networks', Social Sciences, 7/1 (2018), doi: 10.3390/socsci7010013.

Mitchell-Holder, 'Let's talk: effectively communicating with your online students', in Whitney Kilgore, ed., Humanizing Online Teaching and Learning, Chapter 3 (Whitney Kilgore, 2016).

Modic, Dolores, Rončević, Borut, 'Social Topography for Sustainable Innovation Policy: Putting Institutions, Social Networks and Cognitive Frames in Their Place', Comparative Sociology, 17/1 (2018), 100-127, doi: 10.1163/15691330-12341452.

Moon, Suzie, David Birchall, Sadie Williams, Charalambos Vrasidas, 'Developing Design Principles for an e-Learning Programme for SME Managers to Support Accelerated Learning at the Workplace', Journal of Workplace Learning 17/(5/6) (2005), 370–384, Doi: 10.1108/13665620510606788.

Ord, Jon, 'Experiential Learning in Youth Work in the UK: A Return to Dewey', International Journal of Lifelong Education 28/4 (2009), 493–511.

Palloff, Rena and Keith Pratt, 'Collaborating online: Learning together in community', (San Francisco, CA: Jossey-Bass, 2005).

Pathak, Sandip and Priyankiben Vyas, 'E-learning in Modern Digital Environment: A Pragmatic Perspective of Education Institutions', E-learning 6/1 (2019).

Pelet, Jean-Éric, ed., Advanced Web Applications and Progressing E-Learning 2.0 Technologies in Higher Education (Hershey PA, USA: Igi Global, 2019).

Piaget, Jean. T., Psychology of intelligence (Totawa, New Yersey: Littlefield, Adams & Co, 1960).

Poyotos, Fernando, 'Forms and Functions of Nonverbal Communication in the Novel: A New Perspective of the Author-Character-Reader Relationship', Semiotica 21 (1977), 295–338.

Rončević, Borut and Matevž Tomšič, 'Perspectives of information society: bricolage of manifestations', in Rončević, Borut and Matevž Tomšič, eds, Information society and its manifestations: economy, politics, culture (Frankfurt am Main [etc.]: PL Academic Research, 2017).

Rončević, Borut, Raluca Coscodaru and Urška Fric, eds, Go with the Flow: High Performance Computing and Innovations in the Danube Region (London, Budapest, Ljubljana: Vega Press, 2019).

Rončević, Borut, Modic, Dolores, 'Regional Systems of Innovations as Social Fields', Sociologija i prostor: časopis za istraživanje prostornog i sociokulturnog razvoja, 49/191 (2011), 313–333, doi: 10.5673/sip.49.3.3.

Rosenberg, Marc J., E-Learning: Strategies for Delivering Knowledge in the Digital Age (McGraw-Hill Education, 2001).

Sagiv, Lilach, Sonia Roccas, Jan Cieciuch and Shalom H. Schwartz, 'Personal Values in Human Life', Nature Human Behaviour 1 (2017), 630–639.

Singh, Kr Sanjay, Sangrang Brahma, Prasanta Kr Deka, Ibohal Singh, eds, 1st International Conference on Transforming Library 2017 (MRB Publishers, 2017).

Smith, Charlene M., and Todd M. Sodano, 'Integrating Lecture Capture as a Teaching Strategy to Improve Student Presentation Skills Through Self-Assessment', Active Learning in Higher Education 12/3 (2011), 151–62, Doi: 10.1177/1469787411415082.

Snider, Alfred and Schnurer, Maxwell, Many sides: Debate across the curriculum (New York: International Debate Education Association, 2002).

Spante, Maria, Sylvana Sofkova Hashemi, Mona Lundin and Anne Algers, 'Digital Competence and Digital Literacy in Higher Education Research: Systematic Review of Concept Use', Cogent Education 5/1 (2018), Doi: 10.1080/2331186X.2018.1519143.

Stepišnik Perdih, Tjaša, 'Emotional Processing and Relational Marital and Family Therapy', Kairos 12 (2018), 143–163.

Thor, Der, Nan Xiao, Meixun Zheng, Ruidan Ma and Xiao Xi Yu, 'An Interactive Online Approach to Small-Group Student Presentations and Discussions', Adv Physiol Educ 41 (2017), 498–504.

Van Ginkel, Stan, Judith Gulikers, Harm Biemans and Martin Mulder, The impact of the feedback source on developing oral presentation competence, Studies in Higher Education, 2015, Doi: 10.1080/03075079.2015.1117064.

Van Ginkel, Stan, Judith Gulikers, Harm Biemans, Omid Noroozi, Mila Roozen, Tom Bos, Richard van Tilborg, Melanie van Halteren, Martin Mulder, 'Fostering Oral Presentation Competence through a Virtual Reality-Based Task for Delivering Feedback', Computers & Education 134 (2019), 78–97.

Vrasidas, Charalambos and Michalinos Zembylas, 'The Nature of Technology-Mediated Interaction in Globalized Distance Education', International Journal of Training and Development 7/4 (2003), 1–16.

Frane Adam, Maruša Gorišek, Matej Makarovič

# Academic Entrepreneurship in the Framework of Cognitive Mobilisation

**Abstract:** In this chapter, the main focus is on the connection between cognitive (knowledge) mobilisation and high-tech academic entrepreneurship. Cognitive mobilisation refers to an intensive process of generating and utilising/translating knowledge and new ideas into the technological or social/institutional innovations in order to prepare modern societies to becoming sustainable and resilient. This process includes. (1) education (especially on tertiary level – particularly ISCED-8); (2) R&D activities; (3) entrepreneurship in the field of high-tech product and services; and (4) innovativeness (particularly patents). Since no statistical data are available on academic entrepreneurship on European level, we use data on high-tech entrepreneurship as proxy. In our opinion, small and medium-sized high-tech companies are most suitable for examining the sociocultural aspects of knowledge and technology transfer because they combine a market and production orientation with the logic of a research group and its networking with universities/institutes. Using qualitative data – however only in the context of Slovenia as additional case study – we were able to identify mainly market-oriented companies and specifics of so-called academic companies on the other side.

**Keywords:** entrepreneurial university, cognitive mobilisation, small and medium-sized high-tech companies, academic entrepreneurship, spin-offs, innovation

## Introduction

Modern societies face several challenges on their way to becoming sustainable and resilient. In the European context, the Lund Declaration (2009) stresses the need to focus on contemporary Grand Societal Challenges like public health, ageing societies, energy, water, food, climate change, pandemics, and societal security. These challenges are not limited to Europe, but also impact the global, national, regional and local levels around the world. The European Commission (2011, 4) states that quality of life in the future depends on the ability to drive innovation in products, services as well as business and social processes/models.

Such challenges are becoming evident in the transition from an industrial to a knowledge-based mode of production. There is an urgent need for governments and other stakeholders to support the generation of knowledge-based firms, products, technologies, services and a general innovation culture if we are to establish new drivers for long-term knowledge-based economic renewal and growth, and compensate for the private sector's insufficient support for innovation while boosting economic growth (Etzkowitz and Ranga 2009, 799).

Yet, the practice of universities establishing spin-off firms is not so common in continental Europe; in some cases (like Slovenia) there are legal restrictions regarding establishing of firms by public universities[1]. It is, however, possible to form spin-out companies based on researchers holding dual appointments.

Entrepreneurial universities can translate innovative knowledge into new economic and social utility by creating spin-off and start-up companies that result in the diffusion of knowledge, economic activity, and jobs. Academic entrepreneurship is typically seen as being focused on the economic returns from commercialising knowledge produced at the university while social entrepreneurship is viewed more as the endeavours concerned with innovative activities to sustain and create social value (pursuing social, cultural, and environmental goals) that serve community needs and add to social cohesion, employment and reduced inequalities (European Commission 2013). But academic entrepreneurship also holds potential to pursue commercial or social goals, or a mix of both, and has the extra advantage that it can be more proactive in anticipating social problems and produce results in fields apparently not beset by social problems (Cantaragiu 2012, 688)[2].

---

1 Based on its statute, University of Ljubljana can establish a company with the consent of its founder (state) (Statute of University of Ljubljana, article 280), but in practice this consent is never given to public universities in Slovenia, therefore, public universities do not establish spin-off companies in Slovenia.

2 The models of social innovation, such as new strategies, concepts, ideas and organisations that meet social needs and extend and strengthen civil society, although considerable in advancing general societal development, are often neglected in developmental and innovation studies. Social innovation, often stemming from the social sciences and the humanities, tries to find the social value in products and services and is committed to social progress by resolving social issues and meeting social goals where economic resources are a constraint. Knowledge transferred and applied through academic entrepreneurship can promote social innovation and create spiritual and material wealth, help reach social objectives and promote social progress and prosperity (in areas like education, tourism, culture and services). Just as science and technology parks (STP) promote technological and business innovation, social innovation parks, integrated and complementing STPs could promote social innovation, bring a social dimension into innovation and a greater focus on activities with social objectives, as well as enhance general innovation activity with social innovation (Lundstrom and Zhou 2011).

Since we are not only dealing with university spin-offs, and data on academic entrepreneurship in Europe is lacking, we must approach and analyse it from several different perspectives. The latest Eurostat data on high-tech enterprises and certain other indicators recognised as being relevant to the cognitive mobilisation process are used. Our analysis presumes that countries scoring higher in selected indicators of cognitive mobilisation may have more academic enterprises.

The analysis moves to the situation in Eastern and Central Europe, especially in Slovenia where we are researching academic entrepreneurship from the perspective of small and medium-sized high-tech companies, where majority of academic entrepreneurship is happening. In our opinion, small and medium-sized high-tech companies are most suitable for examining the sociocultural aspects of knowledge and technology transfer because they combine a market and production orientation with the logic of a research group. Due to fewer human and financial resources, small firms depend on external knowledge sources and expertise more than larger ones do; to that end often relying on innovation networks with larger firms, innovations supporting or academic institutions (Laperche, Liu 2013), often maintaining closer relationships with the academic sphere and thus making the small enterprise sector a potential market for academic firms (Shane 2004).

In this section, we present the results of qualitative analysis of 17 high-tech enterprises in Slovenia. We identified several companies that could be labelled an academic enterprise by accounting for certain aspects or indicators.

## Cognitive mobilisation in knowledge-based societies

It is undisputed that contemporary societies and economies are characterised by the processes entailed in the production, dissemination and application of knowledge. This means the biggest factors of development and prosperity are education, complex competencies, and the creativity of the R&D sector. The competitiveness of national economies depends ever more on the success of research and innovation systems and the investments made in these systems. In this sense, it is crucial to monitor and correctly measure these systems' performance and dynamics.

However, this represents a one-sided, ideal-type (in M. Weber's terms) view of (post)modern societies. Alongside human and other less tangible forms of capital, economic (financial) capital continues to be important in these societies. Moreover, individual national societies or regions are at various stages of development in which not only evolution but also stagnation and regression remain possible. In the EU, for instance, there are huge differences in terms of innovation capacities and the importance of knowledge in a society. The financial crisis and economic recession are also responsible for the fact that some countries are today paying less attention than before to higher education and science. Especially in the new EU member states and the region of Southern Europe, a brain drain and the impoverishment of human capital are underway. On the other hand, certain countries, especially in Scandinavia, and some others (Germany or Austria) have managed to maintain and even improve their research and innovation potential and systems.

Building on sociological, long-term observations, one may argue that we are dealing with the scientification of society; namely, the ever stronger impact of scientific discoveries and scientific logic on both all social subsystems and everyday life. Yet, we can speak about the socialisation of science that assumes individuals and societies are able to cope with the new risks and opportunities brought by scientific and technological progress. Etzkowitz refers to the intertwining of science, business and policy decision-making institutions as well as civil society (Etzkowitz 2011), while the authors of the Mode 2 production of knowledge stress the "context of application" (Gibbons et al. 1994, also see Carayanis and Campbell 2012).

The major problem of these and similar approaches is that they do not systematically consider the implications of distinguishing commercialisation of knowledge and science from the socialisation or embedding of science and technology within the broader framework of civil society. In this connection, one can apply the concept of cognitive mobilisation which encompasses the response of individuals, organisations and societal subsystems to both the processes of the scientification of society and the socialisation and commercialisation of science and knowledge (Adam 2014). The following accent is important here: cognitive mobilisation not only resides in the capability for action and utilisation of knowledge, but also refers to the ability for (self) reflection, which may trigger long-term and strategic thinking and decision-making.

How to operationalize the term cognitive mobilisation? We proceed from the hypothesis that cognitive mobilisation is an intensive process of generating and utilising knowledge and transferring technological (as well as social and institutional) solutions. This includes: 1) education (especially on the tertiary level – particularly ISCED-8); (2), research or R&D activities; (3) entrepreneurship in the field of high-tech products and services; and (4) innovativeness (particularly patents). In other words, academic entrepreneurship is expected to emerge in societies and settings possessing a sufficient number of highly educated individuals (especially those with a doctoral degree), an adequate level of R&D investments (at least 2.5 % of GDP), above-average patent activities, and where high-tech enterprises are being established.

On the other hand, we can say that academic entrepreneurship is an area where all of the mentioned dimensions of cognitive mobilisation meet and intertwine. Since no data are available on academic entrepreneurship, we use data on high-tech entrepreneurship as a proxy. We also explore qualitative research data concerning academic entrepreneurship, although it is limited to the national context of Slovenia.

## The role of the university – how far in the direction of the commercialisation of knowledge can it go?

In his writings, Etkowitz presents his view on the evolution of the Entrepreneurial University (ENU) as a natural extension of an academic organisation previously concentrated on teaching and research to incorporate technology transfer, firm formation and regional (economic and social) development as its third mission. On one hand, policymakers and political decision-makers are increasingly attracted to the idea of universities directly contributing to national and regional innovation and development. On the other, with significant cuts made to their public funding during the global financial crisis, universities are now forced to not only find additional funding through activities to commercialise knowledge, with which many have been experimenting for at least a decade. Both developments are part of a trend aimed at incorporating elements of the 'third mission' into the structure of the teaching-research model of the university that dominates today.

The ENU mainly originated in the United States, with the MIT and Stanford usually cited as model examples, although various elements of ENU are now found in many universities across the globe. The key issues discussed below are funding sources, IP, cooperation practices, and capabilities and infrastructure.

Most concerns about making universities more commercial and entrepreneurially oriented continue to focus on the potentially detrimental influence of private, especially industrial, funding even though, as Etzkowitz himself notes, similar concerns were voiced when government grants began to be the biggest source of university funding in the USA in the first half of the 20th century. In this view, research could become more short-term and application-oriented and less long-term and basic-research-directed, while publicly funded and Open Access research results might increasingly become the closed-off intellectual property of the industry and private financiers (Gulbrandsen and Slipersaeter

2007, 115). It could be argued that most private investment is channelled into short-term venture investments that promise high returns and are thus fairly secure, whereas more long-term and speculative research possibly able to yield greater societal benefits in the long run goes without funding.

At the ENU, entrepreneurial academics play a dual role by combining academic and business activities, which might lead to serious overburdening. This would require proper administrative support, although during recent cost-rationalisation efforts the trend has been turning more towards giving academics administrative duties on top of all their other academic obligations. Further, such activities require dense networks and cooperation with the business-industry sector to be established, inevitably stretching the personal resources and capacities available[3].

However, irrespective of how important the entrepreneurial university can be for innovation and technology transfer, several challenges and dangers are associated with it as well[4]. The intensive focus on entrepreneurship, gaining profit and marketing research outputs can lead to higher education becoming commercialised. The term "the commercialisation of university sector" (or university research) is not necessarily negative since it can help build university-industry relations and enable products and services to reach consumers. Yet several risks accompany entrepreneurial universities and the extensive commercialisation of higher education. Some authors (Bok 2003; Caulfield and Ogbogu 2015) see it as a threat to academic standards and the academic community. Extensively commercialised higher education can lead to biased findings,

---

3   Perhaps rather than advocating that all academics should strive to also become entrepreneurs, it would be better to form teams of academic individuals, different specialists who jointly carry out academic entrepreneurship activities, instead of trying to combine them within a single person. If such activities are desirable, then they should be rewarded in the academic as well as in the commercial sphere, for example with tenure and promotion, but this also requires proper metrics to be used to measure such output, which should not only be limited to the number of patents and companies.

4   While advances in software and digital technologies generally reduce the costs for start-ups, those that are more dependent on technological hardware for their work still require large investments in laboratories and R&D equipment. Although many universities are now engaging in building the necessary infrastructure in the form of technology parks and incubators, such facilities still require large investments and the conditions offered to new start-ups are not always favourable. In addition, there can be discrepancies in the national legislation, making the mutual ownership of start-ups and spin-offs difficult, hindering the early entrepreneurial process.

subjectivity and damaged integrity as research becomes fuelled by commercial gain and profit. Researchers may feel pressured to develop marketable products, possibly leading to the premature introduction and use of products and services. Research usually takes a longer time, whereas the commercialisation of education promotes fast, successful and useful results. This might compromise the integrity, quality, safety and efficiency of research, or neglect broader and more basic research that has no short-term commercialisation impact. The result could be budget cuts to programmes unable to generate instant revenue and marketable products (Caulfield, Ogbogu 2015).

Regarding this, Philpott et al. (2011) warn that the commercialisation of higher education could create a clear dividing line between different disciplines (especially the traditional humanities and the science- and technology-based disciplines) with respect to how the entrepreneurial university is understood. While some disciplines see it as an opportunity, others view it as a threat to the traditional purposes of university: teaching and research. On one hand, this adds more heat to the internal conflict and, on the other; it could be the cause of funding bias towards different disciplines. The authors therefore argue that the entrepreneurial university ideal is better suited to technologically-oriented universities than comprehensive universities containing numerous standalone faculties (Philpott et al. 2011).

We agree with these critics of the entrepreneurial university and believe that academic entrepreneurship is a more appropriate concept for knowledge transfer. Academic entrepreneurship starts at the university but, once it is established, it only maintains loose ties with it. Entrepreneurial activity mainly occurs outside the university and hence does not create a need for the commercialisation of other university activities.

## Academic entrepreneurship

One can find many different types of enterprises, distinguished by certain aspects. Apart from micro, small, medium-sized, and large enterprises, including multinational corporations, enterprises can also be described according to their technological complexity. We can also differentiate enterprises that are mainly guided by market parameters and those propelled by other circumstances as well. Here, we can mention the regular market enterprises, social enterprises and academic enterprises. Social enterprises aspire to meet the needs of local communities while also focusing on providing employment. Academic enterprises generally follow signals from the market, but also concentrate on the production and cultivation of knowledge. Some authors name other types of enterprises different from classic market-oriented ones. Westlund, for example, also talks about political, civil and innovative enterprises (Westlund 2011).

Academic entrepreneurship (AE) is generally defined in terms of means for transferring knowledge from the university environment to the market through either 'hard' (patenting, licensing, spin-off formation) or 'soft' activities (academic publishing, grant-seeking, contract research). Focusing on the commercialisation, it involves for-profit business creation, university spin-offs, and companies started by academia (Shane 2004; Siegel and Wright 2015). Concerning the role of researchers, we may view AE as the involvement of academic scientists and organisations in commercially relevant activities of different forms, including industry-university collaborations, university-based venture funds, university-based incubator firms, start-ups by academics, and dual appointments of faculty members in firms and academic departments.

The Massachusetts Institute of Technology describes AE as an "innovation-driven entrepreneurship ventures born in academia" (MIT Sloan Executive Education 2014). The multiplicity of definitions and uses of the term may be divided into commercial, knowledge transfers, and value-creation definitions.

Drawing on all three categories, Cantaragiu (2012, 687) defines academic entrepreneurship as a practice performed with the intention to transfer knowledge between the university and the external environment in order to produce economic and social value for both external actors and members of the academia, and in which at least one member of academia maintains a primary role. The academic entrepreneur plays a key role as the initiator of the entrepreneurial practice and must remain its main shaper. Given this definition, AE can develop or transform into either business or social activities, or a mix of both.

Commercial entrepreneurship places emphasis on economic returns, social entrepreneurship on social innovation, by focusing on neglected positive externalities, whereas AE focuses on producing knowledge for the external partners and academia through a meaningful dialogue leading to either commercial or social opportunities. Social entrepreneurship may be seen as an innovative, social-value-creating activity that can occur within or across the non-profit, business or government sectors, and is motivated by a specific problem in the environment. In contrast, AE can be more proactive in anticipating social problems and produce results in fields where there are no apparent social problems.

Academic entrepreneurship is supposed to transform the ways in which knowledge is created, used and diffused, and involves the creation and dissemination of knowledge between the university and the external environment. The faculty, as entrepreneurial thinkers, seek new ways of engaging with the community to create value, developing community-based learning, where the academic perspective is merged with real-life community-based experiences. Communication or dialogue between the knowledge-creation institution of education and the market/society is needed to generate new knowledge and understanding. The engaged university builds bridges with the surrounding community and focuses on community capabilities in order to further its development. It actively participates in the community through engagement, interactivity and interdisciplinary (Cantaragiu 2012).

New technology-based firms are a crucial feature of the modern knowledge economy. They help regional high-tech clusters develop and investments in basic science to be transformed into economic growth, employment and competitive advantages. Governments and other stakeholders are thus interested in creating an environment that is conducive to starting up and nurturing innovative businesses, especially small and medium-sized enterprises, by creating alliances between universities and firms, and commercialising research results through the licensing of intellectual property. New technology-based firms emerge from both well-established industrial firms (corporate spin-offs) and from universities (academic spin-offs).

In modern economies, universities are forced ever more to justify their economic role and demonstrate their impact on society so as to obtain public funding. By appearing to respond to social needs and economic development, universities can enhance their public image, their accountability for public funding, and expand their funding sources to include non-government and public organisations. Entrepreneurial universities have the ability to translate knowledge produced within the university into economic and social utility, and thus regard the creation of new firms as a vehicle for exploiting research results. The formation of university spin-offs is likely to generate more revenue than the licensing of intellectual property, and spin-offs are also important as collaborative partners and future contractors of the university. Academic entrepreneurship is not solely evaluated by the economic returns to the university, but also in terms of wider social and economic benefits like the diffusion of knowledge, new economic activity and contribution to employment.

In terms of the social impacts, engaged universities and (social) academic entrepreneurs need to build bridges and dialogues with their community, focusing on community capabilities, and develop community-based learning where the academic perspective is merged with real-life community (and user) based experiences and needs (Cantaragiu 2012, 687). Knowledge is accordingly not only transferred or translated to address societal problems, but can also be actively steered or shaped to addressing national, regional or local problems and challenges, producing knowledge that is contextually situated and applicable, and build on community strengths to help tackle social challenges and needs; in short, what has been termed socially robust knowledge (Nowotny et al. 2001).

The purpose of this chapter is to analyse academic entrepreneurship in the framework of the transfer of knowledge to new and innovative services and products in Europe. As mentioned, we take a different approach and start by defining academic entrepreneurship separately from the entrepreneurial university. Academic entrepreneurship is defined as a special form of entrepreneurship and contains certain characteristics that distinguish it from classic commercial or, on the other side, social entrepreneurship. We believe the essence of academic entrepreneurship lies not so much in firms established and owned by the university. The accent is on researchers who develop ideas, innovation or prototypes within the location of universities or institutes, and further develop them into commercial products or services outside of the university setting. We also believe the entrepreneurial university concept is less relevant in this context and is more appropriate for universities to leave the entrepreneurial development of ideas to its entities and to only maintain loose ties with them.

## High-tech companies as part of the knowledge-based economy in Europe[5]

The vast majority of academic enterprises are oriented to high-tech, yet only some high-tech enterprises may be called academic. The data we have available relate exclusively to high-tech enterprises.

---

5   According to the Eurostat, three approaches are used to identify technology-intensity – sectoral, product and patent approach. Sectoral approach ranks manufacturing industries using NACE rev. 2 classification. Based on that, high technology includes manufacture of basic pharmaceutical products and preparations, manufacture of computer, electronic and optical products and manufacture of air and spacecraft and related machinery. Product approach is basing on SITC rev. 4, where the following products are defined as high-tech: aerospace, computer-office machines, electronics – telecommunications, pharmacy, scientific instruments, electrical machinery, chemistry, non-electric machinery, armament. According to the patent approach, certain technical fields are defined as high technology: aviation, communication technology, computer and automated business equipment, lasers, microorganism and genetic engineering and semiconductors. OECD uses ISIC rev. 3 classification of manufacturing industries based on their R&D intensities. According to this, high-technology industries include: aircraft and spacecraft, pharmaceuticals, office, accounting and computing machinery, radio, TV and communication equipment, medical, precision and optical instruments industries. These lists are on the first sight somewhat defective,

We begin by examining certain statistical data from Eurostat on the number of high-technology small, medium-sized and large enterprises (HTE) (where we can only identify the small and medium-sized parts of high-tech) and the percentage of workers in this sector. In 2014, there were 1,016,440 units of high-technology industrial and service enterprises in EU member states. The biggest numbers were located in Great Britain and France (namely, more than in Germany). Still, the size of these companies begs a question. In some countries one finds small and medium-sized enterprises, yet elsewhere larger ones are dominant. In Slovenia, there are more than 8,800 HTE, with their numbers growing every year. Per capita, this means Slovenia has have more than Slovakia, Hungary or the Czech Republic, or Austria, Germany, France or the United Kingdom (see Tab. 1 in the Appendix).

Across the entire EU, the HTE sector employs 9,159,900 people. Nevertheless, on average, that amounted to exactly 4 % of all people in the workforce in 2017. The biggest share of employment in these enterprises is seen in Ireland (8.4 %) and Malta, Finland and Slovenia (all reaching 5.7 %), which is more than in Sweden (5.0 %), Austria (4.4 %) or Germany (4.1 %). Greece, Portugal, Lithuania and Romania have the smallest share of employment in HTE (below 3 %).

It is worth adding that Eurostat and EIS identify another category: the percentage of employed in knowledge-intensive activities. These activities require at least one-third of those employed to hold tertiary degrees. The latest EIS data (2018) reveal that, on average, 14.2 % of all workers in the EU were in this category. The highest percentage of employed therein may be found in Ireland and Luxembourg (around 20 %), followed by Great Britain and Sweden (around 18.5 %). In Slovenia, a little less than 14 % of the total employed population is active in this category. Does this then allow us to speak about a knowledge society/ knowledge economy?

The role and relevance of other HTE data is likewise of interest for the analysis, particularly relating to patents and the share of high-technology exports. As far as PCT patent applications are concerned (EIS data), Sweden, Finland, Denmark and Germany have the most (between 6 to 9 per billion GDP). Similarly, based on a Eurostat estimate for 2017 (patent applications to the European Patent Office per million inhabitants), the same countries are at the forefront.

---

for instance, electrical machinery or chemistry are not necessary high-tech, and they could contain medium level of technology.

| | High-tech export | | Number of high-tech enterprises per 1mio inh. | | Employed in high-tech sector | | Employed in knowledge intensive activities | | Graduates at doctoral level | | | Graduates at doctoral level in science, math, comp.... | | | Intramural R&D expenditure | | Number of patents | | Number of high-tech patents | HTE index |
|---|---|---|---|---|---|---|---|---|---|---|---|---|---|---|---|---|---|---|---|---|
| | 2010 | 2018 | 2010 | 2014 | 2010 | 2018 | 2010 | 2017 | 2013 | 2016 | 2017 | 2013 | 2016 | 2017 | 2010 | 2017 | 2010 | 2017* | 2013 | |
| EU 28 | 16,1 | 17,9 | 1.716,95 | 2.303,51 | 3,8 | 4,1 | 35,5 | 36,1 | 1,9 | 2,1 | 2 | 0,8 | 0,8 | 0,7 | 490,9 | 619,9 | 112,82 | 106,84 | 15,775 | |
| Belgium | 8,4 | 10,3 | | 2.610,09 | 4,7 | 4,9 | 41,9 | 43,4 | 1,7 | 2 | 2 | 0,9 | 0,8 | 0,7 | 690,7 | 998,6 | 139,77 | 145,83 | 28,806 | 0,5 |
| Bulgaria | 4,1 | 5,9 | 1.027,11 | 1.426,09 | 3 | 3,9 | 26,3 | 26,9 | 1,2 | 1,5 | 1,5 | 0,4 | 0,4 | 0,5 | 29 | 54,7 | 2,29 | 4,13 | 0,426 | 0,1 |
| Czechia | 16,1 | 17,8 | 3.199,74 | 3.319,88 | 4,3 | 5 | 30,3 | 31,6 | 1,6 | 1,7 | 1,7 | 0,8 | 0,8 | 0,9 | 200,3 | 324,5 | 18,41 | 33,78 | 2,13 | 0,4 |
| Denmark | 9,3 | 9,4 | 2.404,99 | 6.201,98 | 5,5 | 5,2 | 39,5 | 38,2 | 2,9 | 3,2 | 3,2 | 1,3 | 1,4 | 1,3 | 1.281,60 | 1.551,40 | 232,97 | 246,61 | 40,751 | 0,4 |
| Germany | 14 | 15,1 | 1.131,50 | 1.503,04 | 4,3 | 4,2 | 37,6 | 37,1 | 2,7 | 2,8 | 2,7 | 1,2 | 1,2 | 1,2 | 855,9 | 1.200,30 | 286,59 | 228,81 | 27,692 | 0,4 |
| Estonia | 10,4 | 11,5 | 1.924,56 | 3.042,21 | 3,4 | 5,5 | 32,4 | 33,1 | 1,2 | 1,2 | 1,3 | 0,7 | 0,7 | 0,7 | 174,6 | 231,3 | 29,2 | 27,6 | 6,302 | 0,4 |
| Ireland | 19,5 | 34,7 | | | 7,7 | 8,1 | 42,7 | 41,5 | 2,1 | 2,2 | 2,2 | 0,9 | 1 | 0,9 | 586,8 | 646,1 | 71,24 | 77,64 | 18,998 | 1,0 |
| Greece | 5,6 | 4,5 | | 1.134,92 | 2,3 | 2,8 | 32,4 | 35 | 1 | 1,5 | 1,5 | 0,4 | 0,5 | 0,6 | 121,6 | 189,3 | 5,69 | 8,38 | 1,047 | 0,1 |
| Spain | 4,8 | 5,5 | 974,93 | 1.134,84 | 3,4 | 3,6 | 31,7 | 32,5 | 1,6 | 2,6 | 3,7 | 0,7 | 1,2 | 1,5 | 313,8 | 302 | 32,51 | 35,56 | 5,779 | 0,2 |
| France | 20,4 | 20,5 | 1.571,11 | 2.196,24 | 3,9 | 4,1 | 39 | 39,3 | 1,7 | 1,7 | 1,7 | 1,1 | 1,1 | 1,1 | 672,3 | 748,8 | 131,3 | 141,85 | 25,944 | 0,5 |
| Croatia | 7 | 8,1 | 1.321,22 | 1.395,64 | 2,7 | 4,1 | 28,1 | 32 | 1,4 | 1,2 | 1,3 | 0,5 | 0,4 | 0,5 | 77,9 | 101,9 | 7,05 | 4,8 | 0,767 | 0,2 |
| Italy | 6,5 | 7,8 | 1.825,66 | 1.741,24 | 3,3 | 3,5 | 33,3 | 33 | 1,5 | 1,4 | 1,4 | 0,7 | 0,7 | 0,7 | 331,6 | 385,5 | 76,04 | 68,46 | 5,043 | 0,2 |
| Cyprus | 19,3 | 9,5 | | | 2,2 | 3,5 | 34,3 | 38,4 | 0,4 | 0,7 | 0,5 | 0,2 | 0,3 | 0,2 | 105,2 | 127,2 | 9,36 | 10,62 | 2,31 | 0,3 |
| Latvia | 4,8 | 11,2 | 1.441,17 | 2.790,50 | 3,2 | 3,5 | 31,5 | 33 | 1,1 | 0,7 | 0,5 | 0,4 | 0,3 | 0,2 | 51,2 | 70,7 | 7,44 | 11,41 | 7,906 | 0,2 |
| Lithuania | 6 | 7,9 | 644,82 | 1.688,82 | 1,9 | 2,9 | 33 | 32,1 | 1,2 | 0,9 | 0,9 | 0,5 | 0,7 | 0,4 | 69,9 | 131,2 | 5,06 | 7,57 | 1,783 | 0,1 |
| Luxembourg | 30,7 | 7,2 | 3.334,22 | 3.673,05 | 4,1 | 4,2 | 56,6 | 49 | 0,8 | 1,2 | 1 | 0,4 | 0,7 | 1 | 1.202,40 | 1.176 | 152,79 | 93,94 | 9,925 | 0,4 |
| Hungary | 21,8 | 15,6 | 3.554,41 | 3.713,44 | 5 | 5,2 | 34,5 | 33,6 | 0,8 | 1 | 1 | 0,3 | 0,3 | 0,3 | 112,4 | 170,8 | 19,52 | 20,08 | 4,302 | 0,4 |
| Malta | 32,9 | 25,6 | 2.511,91 | | 5,1 | 5,4 | 39,4 | 41,8 | 2,1 | 2,4 | 2,2 | 0,1 | 0,2 | 0,3 | 96,7 | 132,8 | 8,45 | 14,4 | 2,373 | 0,7 |
| Netherlands | 18,6 | 21,3 | 3.259,73 | 5.111,68 | 3,9 | 3,9 | 36,8 | 37,2 | 2 | 1,9 | 2,2 | 0,7 | 0,5 | 0,7 | 657,1 | 859,1 | 184,6 | 203,59 | 34,284 | 0,5 |
| Austria | 11,8 | 13,8 | 2.066,66 | 2.234,31 | 3,7 | 4,2 | 35,8 | 37,1 | 2 | 1,9 | 2,2 | 0,9 | 0,9 | 0,9 | 965,9 | 1.331,30 | 212 | 231,35 | 20,282 | 0,4 |
| Poland | 6 | 8,4 | 1.407,76 | 2.018,55 | 2,7 | 3,2 | 28,5 | 29,5 | 0,6 | 0,6 | 0,5 | 0,2 | 0,2 | 0,2 | 68,6 | 127,3 | 9,5 | 18,08 | 2,307 | 0,1 |
| Portugal | 3 | 4 | 1.427,91 | 1.457,14 | 2,3 | 3 | 28,2 | 31,9 | 1,9 | 2 | 1,8 | 0,8 | 0,8 | 0,7 | 260,8 | 250,7 | 8,99 | 13,8 | 1,54 | 0,1 |
| Romania | 9,8 | 8,4 | 802,82 | 916,11 | 1,8 | 3 | 19,8 | 21,6 | 1,9 | 13,8 | 0,7 | 0,7 | 0,3 | 0,2 | 28,2 | 48,1 | 1,71 | 5,07 | 0,711 | 0,0 |
| Slovenia | 5,3 | 5,8 | 3.242,83 | 4.313,26 | 5,1 | 5,5 | 32,9 | 34,5 | 4 | 2,1 | 1,9 | 1,7 | 3,4 | 0,8 | 364,4 | 387,8 | 51,91 | 55,3 | 4,478 | 0,3 |
| Slovakia | 6,6 | 9,6 | 1.805,24 | 2.747,44 | 3,8 | 4,3 | 30,3 | 31,4 | 2,4 | 2,9 | 2,6 | 0,8 | 1,2 | 0,7 | 77,2 | 137,8 | 8,63 | 10,14 | 1,007 | 0,3 |
| Finland | 10 | 6,1 | 1.645,54 | 1.801,97 | 5,9 | 5,8 | 36,7 | 38,3 | 2,8 | 2,7 | 2,7 | 1,2 | 1,2 | 1,1 | 1.302,70 | 1.121,70 | 260,13 | 235,68 | 57,65 | 0,4 |
| Sweden | 14,5 | 11,3 | 5.382,80 | 5.583,39 | 4,8 | 5,3 | 42,6 | 45,1 | 2,8 | 2,7 | 2,6 | 1,4 | 1,4 | 1,3 | 1.270,80 | 1.615 | 301,55 | 283,46 | 51,412 | 0,5 |
| United Kingdom | 17,7 | 16,7 | 2.333,19 | 2.902,22 | 4,2 | 4,8 | 43,2 | 43,7 | 3 | 3,1 | 3,1 | 1,4 | 1,4 | 1,5 | 491,6 | 591,1 | 85,64 | 82,62 | 15,759 | 0,6 |

**Tab. 1:** Eurostat data on selected indicators of the innovative economy. Source: Eurostat (2019): Science, technology and innovation database. *HTE Index: Own calculation based on the same Eurostat data. Available at: https://ec.europa.eu/eurostat/web/science-technology-innovation/data/database (Accessed in September 2019)*

Slovenia is estimated to have 55,30 patent applications per million inhabitants, scoring below the EU average, but still performing better than the Czech Republic, Hungary or Slovakia.

The data on high-technology patent approvals, however, do not match the data on the share of high-technology products exported (per country or in EU total). At the EU level, this share was 17.8 % in 2017. It is intriguing that those countries with the highest number of patents generally have low export shares (Great Britain is an exception). This is true for Finland, Sweden, Denmark and Germany, putting them all below the EU average. Finland even has half the average percentage share (6.6 %). Malta (25.5 %), as well as Luxembourg and Ireland, have the highest export rate of high-technology products. Malta is especially interesting from a methodological standpoint. A certain elaboration was expected, but the available publications – such as IUS – do not provide one.

The case of Slovenia is likewise fascinating. It has a very small share of high-technology exports (5.5 %). Still, publications and score boards using a combination of high- and medium-technology exports place it at the very top.

For instance, the publication Research and Innovation Performance in EU Member States and Associated Countries states: "regarding its medium and high-tech trade specialisation /.../ it is second only to Germany" (European Commission, 2013: 250). The adequacy and comparability of the statistical data in Eurostat (and IUS) may, however, be questioned. This statement relies on the indicator "contribution medium and high-tech (MHT) products to trade balance" which elevates Slovenia 105 % above the EU average (IUS, 2014: 66). Interestingly, the majority of countries are placed around the EU average. Even more surprising is that Macedonia – which according to all other development and innovation indicators expectedly occupies the bottom among European states – has the exact same value (105 % of the EU average).

According to this overview of the statistical and indicator databases, we may conclude that the European Union has yet to reach the point when it can be called a knowledge economy. High-technology enterprises are still on the brink, their role is not a determinant of a competitive economy. On the other hand, we are dealing with certain unexplainable, inconsistent, even contradictory findings and data. The databases require a breath of fresh air. Corrections are overdue for Eurostat, as is better coordination with national statistical offices. The current condition is unbearable because it guarantees neither comparability between states nor temporal comparisons.

Studies show the EU is successful in the production of high-technology knowledge but lags behind other regions, particularly the USA, in its commercialisation within the key high-technology domains of biotechnology (Ernst and

Young, 2013: 25–33), nanotechnology (Savolainen, 2013: 73) and advanced ICT (van Lieshout et al., 2008: 158). Solely the production of high-quality knowledge without an appropriate support system, incentives and investments is unable to ensure the transfer into new products and processes, and their commercialisation – hence contributing to developing the economy and society while simultaneously addressing various challenges and needs.

## Hierarchical cluster analysis

In addition, we briefly reviewed the country patterns (based on the latest Eurostat data (Eurostat, 2019) in terms of EU member states taking account of the following measures of cognitive mobilisation:

High-tech entrepreneurship

- high-tech export (as a share of exports of all high technology products in all exports in 2017)
- employed in the high-tech sector (per 1 million inhabitants)
- employed in knowledge-intensive industries (as a share of total employment)

Tertiary education

- all new doctors of science in 2017 (regardless of their age) as a percent of the population aged 25–34

Performance of R&D sector

- intramural R&D expenditure (in EUR per inhabitant)

Innovativeness

- number of patents (by priority year per million inhabitants)

For these variables, we applied hierarchical cluster analysis (average linkage method) to identify the groupings on this basis. This enables us to distinguish groups of countries where cognitive mobilisation has been present for longer and where academic entrepreneurship is more likely to develop. The results are shown in a dendrogram (Fig. 1).

In this regard, we can observe a rough division of most EU countries into the: (1) North-Western Europe block, stronger in cognitive mobilisation (including all old EU members, except the Mediterranean); and the (2) Mediterranean-East-Central Europe, weaker in cognitive mobilisation (including all post-communist and Mediterranean countries).

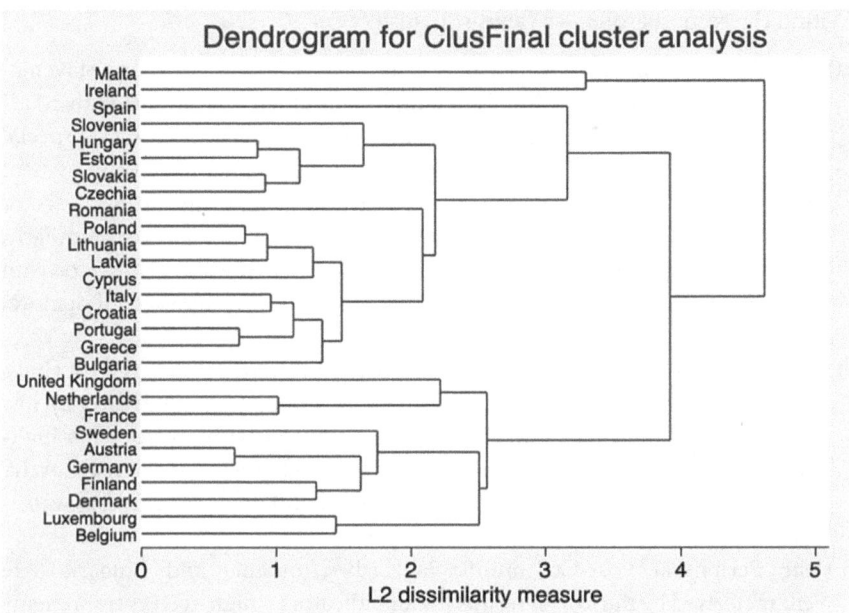

**Fig. 1:** Clusters of EU countries in terms of high-tech export, employment, doctoral graduates, R&D expenditure and patents. Source: Authors

The first group may be subdivided into:

a) Mixed group: Nordic-Central-European-Benelux (Finland, Sweden, Austria, Germany, Denmark, Luxembourg and Belgium), scoring high levels for innovativeness and performance of the R&D sector (heavy investment of their own in R&D and a correspondingly very high number of patents. They also score highly for high-tech entrepreneurship (high employment in the HT sector and knowledge-intensive industries, but mostly a below-average level of HT exports). Further, they have mostly score highly in tertiary education (mostly a high proportion of doctoral graduates).

b) the Western-European block (UK, Netherlands, France): compared to the first group, they score highly for high-tech entrepreneurship (they are even more oriented to high-tech production and corresponding high-tech exports, but have medium or high levels of HT and knowledge-intensive employment). Their level of R&D performance is high, as is their level of innovativeness. Regarding the tertiary education indicator, their proportions of doctoral level graduates vary from medium (France) to very high (UK).

Within the second group, we can distinguish:

a) the East-Central European block (Hungary, Czechia, Estonia, Slovenia) which score moderately highly for high-tech entrepreneurship with moderately high employment in HT enterprises and (except Slovenia, a special case in this sense) high levels of high-tech exports, but with less employment in knowledge-intensive activities. Similarly, they lag behind in R&D sector performance with low R&D expenditure and in innovativeness. Similarly, they score at medium or low levels for the tertiary education indicator with a medium or, in the cases of Estonia and Hungary, low numbers of doctoral graduates.

b) the Mediterranean and Post-Communist semi-peripheral laggards (Italy, Poland, Lithuania, Latvia, Croatia, Portugal, Greece) characterised by low scores for the high-tech entrepreneurship indicator with low levels of high-tech exports, employment in HT enterprises and knowledge-intensive activities. They also score low on R&D performance and innovativeness, as well as on tertiary education.

c) the Peripheral Post-Communist laggards (Romania and Bulgaria) are characterised by the lowest results for all indicators – high-tech entrepreneurship, tertiary education, performance of the R&D sector, and innovativeness. While Bulgaria is somewhat closer to the Mediterranean countries, Romania is the most extreme case in terms of lagging behind the others.

We should also mention three special cases that do not belong to any of these groups:

• Ireland with extremely high scores for high-tech entrepreneurship, high scores for tertiary education and medium ones for innovativeness. It differs from the North-Western group by its extremely high levels of HT employment and exports.

• Malta with very high levels of high-tech entrepreneurship, but a very low level of R&D performance, innovativeness and tertiary education indicators. It is more similar to the North-Western group in terms of high HT exports and employment, but closer to the Mediterranean-East-Central group in terms of R&D expenditure and patents.

• Spain is closer to the Mediterranean-East-Central European pattern but stands out for its very high number of doctoral graduates[6].

---

6   Regarding the number of doctorates, Eurostat presents the number of new doctorates on annual level. It is not known why Spain had such high number of new doctorates

## High-tech entrepreneurship index

As mentioned, high-tech entrepreneurship is used as a proxy indicator of academic entrepreneurship because no data are available on the number of academic enterprises in Europe. Since we observe high tech-entrepreneurship as a single dimension in conceptual terms, combining HT exports, employed in HT sector and employed in knowledge-intensive industries, we may combine them into a single index, which could be called the High Tech Entrepreneurship Index (HTEI).

We should note that high-tech entrepreneurship measured in this way not only involves the activities of domestic entrepreneurs but also of foreign ones while operating through foreign or multi-national companies operating in a given country. This is especially visible in HT exports, which are obviously not generated just by domestic but also by foreign/multinational companies.

The index is constructed as a dimension index ranging from 0 (the lowest level of HT entrepreneurship among our countries) and 1 (the highest level) derived from the sum of the standardised values of:

- high-tech exports (as the share of exports of all high-technology products in all exports in 2017)
- employed in the high-tech sector (per 1 million inhabitants)
- employed in knowledge-intensive industries (as a share of total employment)

We are interested in how much entrepreneurship may be regarded as the result of tertiary education, performance of the R&D sector, and innovativeness. The relationships are not sufficiently straightforward to develop a regression model with statistically significant coefficients, but we can calculate bivariate Pearson correlation coefficients and graphically present the relationships, as seen in Fig. 2.

Tertiary education is moderately related to HT entrepreneurship (Pearson $r = 0.33$). Special cases are particularly interesting. On one hand, the number of doctorates in science awarded in Spain is not reflected in its entrepreneurship. On the other hand, the significantly foreign-driven entrepreneurship of Ireland and Malta may flourish despite their comparatively lower achievements in the higher education dimension. For more info see Fig. 3.

---

in 2017 (and in 2016). In contrast, OECD data shows the total number of persons with doctorate in an entire period until the last year (2018). Here, we can see that Spain belongs to countries with the smallest number of doctorate holders per 1000 people between 25 and 34 years old.

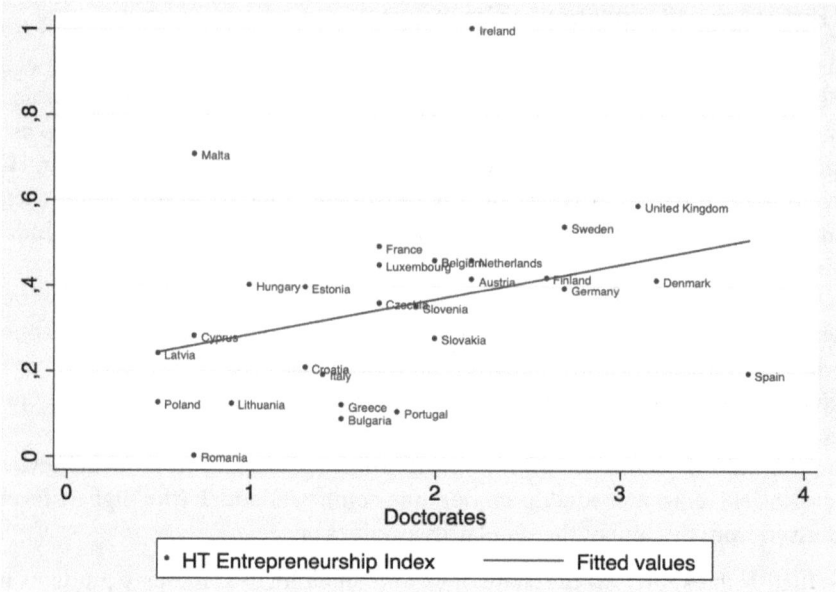

**Fig. 2:** Relationship between HT entrepreneurship and tertiary education (doctoral degrees). Source: Authors

We can also observe a relationship with a higher correlation ($r = 0.47$) between the performance of the R&D sector and HT entrepreneurship. Again, we can confirm the relevance of this sector for HT entrepreneurship but accompanied with the note that it could also be based on the R&D generated abroad, as perhaps best seen with Ireland and Malta. For more info see Fig. 4.

## High-tech firms and academic entrepreneurship in East-Central Europe and Slovenia

Proceeding from the hierarchical cluster analysis, we can conclude that academic entrepreneurship is less developed in East-Central Europe due to the lack of an appropriate environment in the cognitive mobilisation sense. Even though some elements of entrepreneurship are more developed in these countries (e.g. high-tech exports in some countries, or the share of the employed in the high-tech sector in Slovenia), these countries score lowly for innovativeness and R&D sector performance. At the same time, it seems that the high levels of high-tech

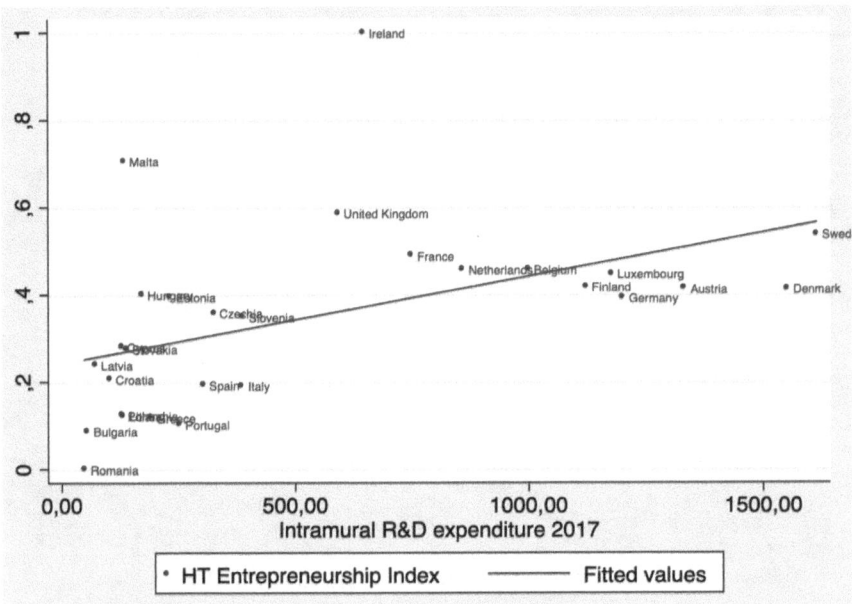

**Fig. 3:** Relationship between HT entrepreneurship and the performance of R&D sector.
Source: Authors

exports are only the result of foreign capital and the presence of international high-tech firms in these countries.

Tchalakov, Mitev and Petrov (2010) see the emergence of academic entrepreneurship in Eastern Europe as a bottom-up spontaneous phenomenon that laid the grounds for rebuilding certain economic sectors in transition. As they state, there are several characteristics of Eastern European spin-offs that differ from the general characteristics of academic entrepreneurship and they notice two conflicting tendencies in the post-socialist spin-offs: the authentic form of academic entrepreneurship and a specific rent-seeking strategy to drain valuable public assets. Academic spin-offs in Eastern Europe not only transfer knowledge to industry but often play the role of an intermediary between emerging private businesses and foreign companies. As some sectors of the economy have collapsed, academic spin-offs have also often acted as the sole providers of technical services in Eastern Europe. However, academic spin-offs do not vary significantly from the rest of the economy in Eastern Europe in transition in that they are both characterised by low business and management skills.

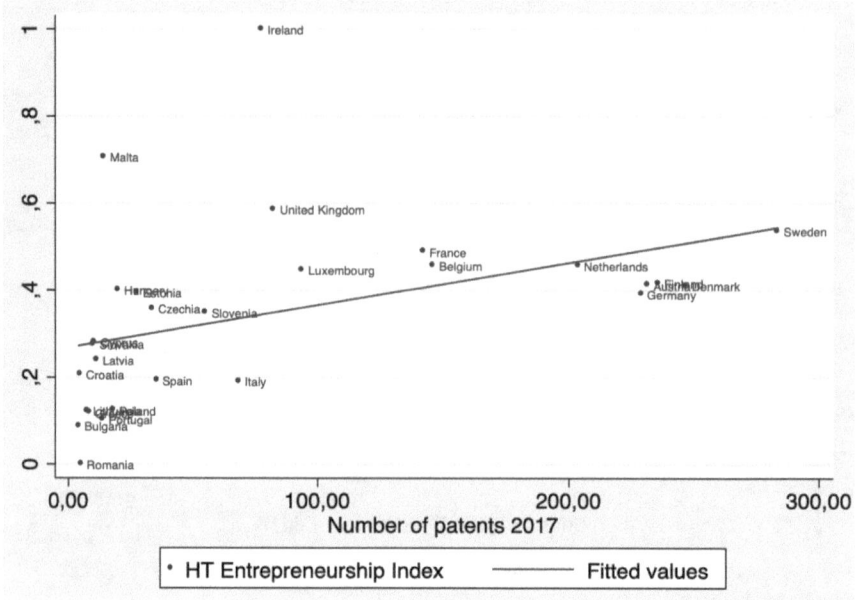

**Fig. 4:** Relationship between HT entrepreneurship and innovativeness. Innovativeness also turns out to be mostly significant for HT entrepreneurship with a correlation coefficient of 0.42. Again, Ireland and Malta stand out for the reasons explained above. Source: Authors

Studies of academic spin-offs from Estonia and the Czech Republic similarly show a lack of entrepreneurial spirit and experience. In their research, Peterkova and Wozniakova (2015) notice there are many business incubators in the Czech Republic, but not many start-up or spin-off firms. They connect this with the lack of shared entrepreneurial experiences and spirit and an unwillingness to take risks (Peterkova and Wozniakova 2015).

In Estonia, the economy, especially exports, was dominated by labour-intensive sectors and low value-added products that resulted in low productivity and interest in entrepreneurship. Universities started to more actively promote and support scientists with the commercialisation of research outputs after 2000 through different grants. Yet, research shows that scientists and researchers need more education in business management skills, and more favourable economic and institutional development towards small high-tech companies is also required (Andrijevskaja et al. 2006).

Moreover, some Eastern European countries like Romania deal with large regional differences in innovation setting due to a very unbalanced distribution of universities, lack of innovation and entrepreneurship tradition and trust, producing little interest in the R&D activities of companies (Serbanica 2011).

When looking more closely at the case of Slovenia, strictly speaking, the restrictive legislation means that actually there are no spin-offs formed directly by universities and public institutes. Academic entrepreneurship in Slovenia emerges in the form of university spin-outs based on dual appointments of academics where the firms are not owned by the universities or institutes, but by the researchers or academics who transfer knowledge and commercialise their research outcomes created in their work at universities and institutes by establishing firms. In most cases, academics remain fully employed at their university or institute and take on an additional 20 % employment in private firms.

Therefore, certain small and medium-sized companies keep close ties to academia and may be characterised as academic enterprises given their organisational structure. Further, in recent years a strong network of innovation supportive mechanisms has developed in the Slovenian academic and business sphere specifically targeted at the promotion of knowledge transfer, entrepreneurship and start-up formation. In 2018, the Slovenian government introduced the action plan "Slovenia – a land of start-ups" with which it plans to further promote such activities and remove barriers to the creation of start-ups[7] (Ministry of Economic Development and Technology 2018).

---

7 At the university level, there are several mechanisms supporting academic entrepreneurship, especially in public universities in Slovenia. The most developed in this sense is the University of Ljubljana with a university incubator (*Ljubljanski univerzitetni inkubator*) which was established in 2008. Besides the incubator, public universities in Slovenia offer several different innovation and entrepreneurship supporting services, such as knowledge-transfer offices, innovation institutes and technology centres. Further, there are three technology parks in Slovenia all owned by the municipality and largely involved in a general national level network scheme of innovative supportive initiatives. There are also 14 local public business incubators (owned by a municipality or a group of municipalities and supported by various government or EU-level schemes) and a few private business incubators (e.g. the incubator of the private Gea College). Business incubators were mainly created after the year 2000 and created from 10 to 40 companies on average. The biggest innovation supportive mechanism in Slovenia, the Startup initiative is the result of the cooperation of a network, especially the University of Maribor's incubator Tovarna Podjemov and the Ljubljana Technology Park. The initiative joins a university incubator, 3 technology parks, 4 local incubators

## Organisational culture in small and medium-sized high-tech companies

We build on the recognition that small and medium-sized high-tech companies (SMHTCs) are the best suited or at least the most grateful milieu, in a research sense, for the study of the sociocultural aspects of knowledge and technology transfer. These companies combine a market and production orientation with the logic of a research group. In this sense, we can talk of a symbiosis of two cultures, that is, of a classical business-entrepreneurship and an epistemological-academic culture. They need to develop both a strategy of business success as well as a strategy of knowledge management and cognitive mobilisation. To some extent, this is true of many other companies dealing with more sophisticated technologies, but SMHTCs are in this regard particularly exposed and forced to effectively coordinate these two cultures and strategies. They need to take market signals into account as well as new ideas and discoveries in scientific and technological innovations. Further, they must establish numerous contacts and links both with companies and actors in the business environment and with competent state institutions, as well as the academic sphere (universities and institutes).

On the other hand, some companies remain more committed to the logic of research groups and combine that with business functions and a market orientation. In this case, some authors use the term academic entrepreneurship. Yet not all SMHTCs fall into this type of entrepreneurial orientation, primarily those companies formed by way of spin-off companies under the auspices of universities.

In this chapter, we include data and findings from our own research (Adam, et al., 2014) in which we conducted 11 interviews with representatives of small and medium-sized high-tech companies in Slovenia[8].We complement this with data from similar research conducted in 2016 (Gojkovič 2017), where six other interviews were conducted based on the same questionnaire. These studies shown that certain SMHTCs chiefly pursue a market and classic entrepreneurial orientation which they combine with R&D activity. This is especially seen where the main funding comes from sales while research and development follows market needs and demands (Gojkovič 2017).

---

and 5 government agencies in creating over 340 events a year for over 400 start-up companies.

8  Data from this study are archived and publicly available from the Social Science Data Archive at: https://www.adp.fdv.uni-lj.si/opisi/vtpod14/ (August 2019).

Our study also identified some companies that may be counted among academic enterprises. However, we also took several additional aspects or indicators into account, such as the number of employees with doctoral degrees, the involvement of companies in national and international research projects, publication activity of company members or the ITS research group, and the frequency of contacts with domestic and foreign academic institutions such as co-authorship of patent applications or collaboration in centres of excellence. As noted, in Slovenia no spin-offs are formed directly by universities and public institutes (due to restrictive legislation, although the new law on higher education should bring some changes in this area). Yet there are spin-outs and several cases of academics and professors holding dual appointments. Therefore, one finds SMHTCs with a scientific character that are well embedded in the academic environment. Further, some companies produce products or services whose customers are principally scientific organisations (such as measuring devices or components of such devices, or software tools used for research purposes).

The hybrid structure of these companies is reflected in their organisational structure and culture, and their social capital. In this regard, our current research has yielded certain findings we would like to verify and upgrade in the coming years. We also intend to improve the relevance and representativeness of sampling.

Pustovrh et al. (2017) estimated the total number of 105 SMHTC in the category of high-tech manufacturing and 2,156 in the category of knowledge intensive services in Slovenia in their study from 2014. Based on a survey of 105 SMHTC representatives, they concluded that Slovenian SMHTCs are unlikely to engage in Open innovation information exchange activities such as cooperation with knowledge transfer offices, entrepreneurship incubators and research partners; scientific publishing or conference attendance. They also noticed that Slovenian SMHTCs are unlikely to engage in Open innovation collaboration where the authors, among others, included cooperation with commercial labs, private R&D institutions or universities.

High-tech companies, both small and large, form the backbone of an economy based on knowledge and innovation. This part of the economy is supposed to be the most propulsive and combine knowledge from academic institutions with entrepreneurial ideas on how to commercialise such knowledge in the form of innovations and advanced technologies. However, this segment still accounts for a small share of economic activities in the European Union. According to Eurostat data (2017), this part of the economy employs 4 % of all employees. If we add to this the share of employees in business activities that are knowledge-intensive – these activities are defined as those with at least one-third of employees with a tertiary education – then this share amounts to 14 %. Slovenia is supposed to be slightly above the EU average in this regard, in terms of both the number of companies and percentage of employees but is distinctly below the average in terms of exports of high-tech products and high-tech patents. This holds certain implications for the cooperation among SMHTCs, the joint exploitation and integration of research, and development potential.

Given our theoretical starting points and the results of the qualitative study relying on interviews, it may be concluded that important differences exist in the organisational culture between academic and regular market-oriented enterprises. Instead of rules and procedures corresponding to a hierarchy, academic companies typically have a small number of organisational units and management levels. Research and development activities are organised in the form of project groups that allow for the employees to move between these groups. Employees also participate in important business decisions, particularly in companies of which they are also co-owners. Therefore, the organisational structures of the examined academic companies are oriented towards the organisation's flexibility and adaptability. The flat hierarchy allows important decisions to be made quickly, permitting these companies to respond better to sudden changes in the business environment than regular market-oriented companies.

Communication in academic companies is mainly based on two-way social interaction, with regular meetings enabling the employees to compare their viewpoints in order to reach solutions and achieve common goals. The flat hierarchy allows them to have open and informal communication with their superiors. These companies' tendency to resolve conflicts by comparing arguments and seeking consensus supports the finding that communication in academic companies runs according to the spirit of democratic discourse which, rather than giving in to pressures regarding the choice of interlocutor or the contents of the conversation, asserts the power of argumentation and consensus-seeking.

Here it should be noted that important sources of new knowledge for academic companies include programme groups and research projects, where the companies collaborate with universities and institutes. With respect to the inclusion of new members, academic companies are prepared to invest more in social capital, largely through acquiring new knowledge and practices. An example of this is the inclusion of foreign scientists in research groups, which increases the impact of their scientific publications as well as the international recognisability of the company. With errors at work being considered a means of learning, experiential learning based on personal experience is encouraged and the employees are expected to be able to absorb knowledge and integrate experience with their own existing knowledge. In contrast, market-oriented enterprises tend to devote pay attention to formal education and foreground past experience in an attempt to reduce potential errors in their future work.

In terms of business environment, all of the interviewed companies largely take the business environment in Slovenia for what it is, and try to operate within this framework. The Supervisor data shows that all companies receive some income from the government sector. Standing out the most are academic companies which, compared to market-oriented ones, more often depend on state resources. Among other findings, there is a clear pattern showing that the companies associated with foreign partners in terms of ownership create most of their profits abroad.

Analysis of the companies' ownership structure reveals that in most cases the employees are not co-owners of the firms. Moreover, while academic companies are generally under Slovenian ownership, all of the market-oriented firms are (co)owned by foreign companies or citizens. Academic companies are found to be more active in terms of collaboration with other companies and the academic environment, thus acknowledging the integrative role played by social capital in access to new knowledge and information.

Academic companies were established on the basis of (preliminary) research work in the academic environment. This study shows that the companies' co-founders transferred the overall patterns of the organisational culture existing in the academic environment to their new organisations, as evident from: (1) their strong connections with academic institutions; (2) flexible organisational structure, based largely on project groups and teamwork; (3) open and informal communication based on democratic discourse; (4) the transfer of knowledge based on planned and formal co-operation, within programme and research groups, with other companies and institutions from the academic environment; (5) the inclusion of new members seen as the strengthening of the company's social capital and its ensuing innovation capacity, while the importance of the formal education acquired is in the foreground; (6) their greater adaptability to the business environment in Slovenia; (7) a considerable share of these firm's incomes deriving from state sources; and (8) more intensive integration with the academic environment and other companies in terms of establishing affiliated business entities.

Based on all of this, one can define the cultural framework in which academic entrepreneurship is establishing itself. Undoubtedly, the academic entrepreneurship concept also represents links between the academic sphere and external environment where the market aspect with its logics of profit intertwines with the desire to transfer knowledge and create new value that is not necessarily only economic. While on one hand market-oriented SMHT enterprises have much in common with academic companies, they also reveal certain patterns of organisational culture that distinguish them: (1) their business strategy reduces their manoeuvring space while making risky decisions; (2) their communication is mainly formal; (3) they co-operate less with the academic sphere; (4) they prioritise practical experience over formal education; (5) state resources do not play a major role in their operations; and (6) their partner networks are important, but less extensive and varied (Gojkovič, 2017).

## Conclusion

With respect to the wider context, we may conclude that the EU has not yet developed the necessary critical mass to make a breakthrough in terms of cognitive mobilisation. In this text, we operationalized cognitive mobilisation with four dimensions:– education, high-tech entrepreneurship, research and R&D activities, and innovativeness. Of course, it is very risky speaking about the EU in general given the huge differences seen between the members. However, even a close examination of the countries that are the most profiled regarding cognitive mobilisation does not allow us to conclude that a distinctive turn has been made towards the more intensive use of knowledge and innovation-related activities, despite clear trends in this direction.

We analysed recent data from Eurostat to distinguish EU countries with the highest levels of cognitive mobilisation, meaning that academic entrepreneurship is most likely to emerge in those settings. We performed a hierarchical cluster analysis that identified two bigger groups of countries based on their cognitive mobilisation level. The level of cognitive mobilisation based on the four dimensions that were described is highest in Nordic and some Central European countries, such as Finland, Sweden, Austria, Germany, Denmark, Luxembourg and Belgium. Therefore, we expect academic entrepreneurship to be most developed in those countries. On the contrary, the level of cognitive mobilisation was lowest in some Eastern European countries like Bulgaria and Romania, which means there is a smaller chance of academic entrepreneurship emerging there.

Since no data were available on academic entrepreneurship in Europe, we used high-tech entrepreneurship to proxy for academic entrepreneurship. For the purpose of our analysis, we created the High-Tech Entrepreneurship Index for each country. We analysed to what extent entrepreneurship could be seen as the result of tertiary education, performance of the R&D sector, and innovativeness. Although the relationships are not sufficiently straightforward to develop a regression model with statistically significant coefficients, we were able to graphically present the relationships. We can observe that all three dimensions are crucial for the emergence of high-tech entrepreneurship and are therefore important factors in the development of academic entrepreneurship.

It is important to note the problem of the adequacy and accuracy of the measurement and classification of individual activities. This was emphasised with reference to Eurostat's data and the data-acquisition methodology used by the Slovenian Statistical Office (Adam, 2014) to establish the scope, sources and sectors pertaining to the research and development activities.

In further analysis, we focused on the micro level and the case of Slovenia where we presented the results of our qualitative analysis of the organisational culture, the networking method, and the overall mode of SMHTE operations (for our sample of these enterprises) in Slovenia. The results show they can be reasonably divided into mainly market-oriented companies and so-called academic companies or companies belonging within academic entrepreneurship. All SMHTCs have a hybrid structure, intertwining the business-market and research-epistemological functions, with the latter aspect being more developed and sometimes even foregrounded in academic entrepreneurship. The question is whether these companies can be seen as the most propulsive expression of new forms of entrepreneurship and new organisational culture. We do not know whether the category of academic companies is in a developmental phase that will later be surpassed, or if this is a more lasting organisational structure. Yet, the fact remains that the companies studied in this work show rapid changes in ownership, in establishing new business entities, and in their internationalisation.

Another evident fact is the huge influence of the state or the way companies respond to the schemes set up for the funding of research and development activities in companies, or regarding other grants awarded by the state. Moreover, these (academic) companies are included in the centres of excellence, competence centres and other similar forms of co-operation between the state, business sector and academic sphere. We also do not have enough knowledge about the efficiency of these forms and their experiences. Another dilemma is whether (some of) these companies would even be able to survive without external (state) help or access to EU resources. For example, the website of one of the best known SMHTE in Slovenia, Zemanta, claims that, rather than direct state aid, what is needed are lower labour costs and taxes (see Adam 2017).

# Appendix

## Selected indicators

- High-tech exports: Exports of high-technology products as a share of total exports in percentage
- Number of high-tech enterprises: Enterprises in high-tech sectors by NACE Rev.2 activity in number

- Number of high-tech enterprises per 1 million inhabitants: calculated per million inhabitants
- Employed in high-tech sector: Employment in technology and knowledge-intensive sectors at the national level, by sex (from 2008 onwards, NACE Rev. 2) in percentage of total employment: High-technology sectors
- Employed in knowledge-intensive activities: Annual data on employment in knowledge-intensive activities at the national level, by sex (from 2008 onwards, NACE Rev. 2) in percentage of total employment: Total knowledge-intensive activities
- Graduates at doctoral level: All graduates at doctoral level by sex and age group per 1000 of population aged 25–34
- Graduates at doctoral level, in science, maths, computing, engineering, manufacturing, construction: by sex per 1000 of population aged 25–34
- Intramural R&D expenditure: Intramural R&D expenditure (GERD) by sectors of performance in all sectors: in Euro per inhabitant
- Number of patents: Patent applications to the EPO by priority year per million inhabitants
- Number of high-tech patents: High-tech patent applications to the EPO by priority year: per million inhabitants

## References

Adam, F. (2014). Measuring National Innovation Performance. Heidelberg and New York: Springer V.

Adam, F. (ed.) (2017). Slovenia Social, Economic and Environmental Issues. New York: Nova Science Publishers.

Adam, F., and Westlund, H. (2013). 'The Meaning and Importance of Socio-Cultural Context for Innovation Performance'. In Adam, F., and Westlund, H. (eds), Innovation in Socio-Cultural Context. Vol. 84, pp. 1–21. New York: Routledge.

Adam, F., Gojkovič, U., Hafner, A., Pustovrh, T., and Zajc J. (2014). Visokotehnološka podjetja – vpliv organizacijske kulture in socialnih omrežij na prenos znanja. Ljubljana: IRSA.

American Association for the Advancement of Science. (2014). 'The Third Way: Becoming an Academic Entrepreneur', Science Mag, <https://www.sciencemag.org/careers/2014/03/third-way-becoming-academic-entrepreneur>, accessed 17 July 2019.

Andrijevskaja, J., Mets, T., and Varblane, U. (2006). 'University spin-off processes: Creating knowledge-based entrepreneurship in Estonia'. at the 14th Nordic Conference on Small Business Research, May 2006 (pp. 11–13).

Balazs, K. (1996). Academic Entrepreneurs and their Role in 'Knowledge' Transfer. ESRC Centre in Science, Technology and Environment Policy.

Bok, D. (2003). Universities in the Marketplace. The commercialization of higher education. Princeton: Princeton University Press.

Cantaragiu, R. (2012). 'Towards a conceptual delimitation of academic entrepreneurship', Management & Marketing, 7 (4), 683–700.

Carayannis, E. G., and Campbell, D. F. (2012). Mode 3 knowledge production in quadruple helix innovation systems. New York: Springer.

Caulfield, T., and Ogbogu, U. (2015). 'The commercialization of university-based research: Balancing risks and benefits', BMC medical ethics. 16 (1), 70.

Ernst and Young, (2013). 'Ernst and Young Attractiveness Survey: UK 2013' <https://www.ey.com/Publication/vwLUAssets/European-Attractiveness-Survey-2013/$FILE/European-Attractiveness-Survey-2013.pdf>, accessed 22 August 2019.

Etzkowitz, H. (2011). 'The triple helix: science, technology and the entrepreneurial spirit', Journal of knowledge-based innovation in China, 3 (2), 76–90.

Etzkowitz, H., and Ranga, M. (2009). 'A trans-Keynesian vision of innovation for the contemporary economic crisis: 'picking winners' revisited', Science and Public Policy, 36 (10), 799–808.

European Commission. (2011). Europe 2020 Flagship Initiative Innovation Union. Directorate-General for Research and Innovation. <https://ec.europa.eu/research/innovation-union/pdf/innovation-union-communication-brochure_en.pdf> accessed 7 July 2019.

European Commission. (2013). Entrepreneurship 2020 Action Plan. Reigniting the entrepreneurial spirit in Europe. <https://eur-lex.europa.eu/legal-content/EN/TXT/PDF/?uri=CELEX:52012DC0795&from=EN> accessed 7 July 2019.

Eurostat. (2019). 'Science, Technology and Innovation Database. <https://ec.europa.eu/eurostat/web/science-technology-innovation/data/database> accessed 12 September 2019.

Frissen, V., and van Lieshout, M. (2006). 'ICT in everyday life'. In Slob, A. and Verbeek P. P. User Behavior and Technology Development, pp. 253–262. Dordrecht: Springer.

Gibbons, M., Limoges, C., Nowotny, H., Schwartzman, S., Scott, P., and Trow, M. (1994). The New Production of Knowledge: The Dynamics of Science and Research in Contemporary Societies. London: SAGE Publications.

Gojkovič, U. 2017. Sociokulturne razsežnosti prenosa znanja v okviru akademskega podjetništva. Nova Gorica: Fakulteta za uporabne družbene študije, doctoral dissertation.

Gulbrandsen, M., and Slipersaeter, S. (2007). 'The third mission and the entrepreneurial university model'. In Bonaccorsi, A. and Daraio, C. (eds.), Universities and strategic knowledge creation, pp. 112–143 Cheltenham, Northampton: Edward Elgar Publishing.

Gümüsay, A. A. (2011). 'Socio-Academic Entrepreneurship', Gümüsay, <http://guemuesay.com/2011/07/07/socio-academic-entrepreneurship-iv/> accessed 7 July 2019.

IGI Global (2019). 'Disseminator of Knowledge: What is Academic Entrepreneurship?', <https://www.igi-global.com/dictionary/entrepreneurial-mission-of-an-academic-creative-incubator/47148> accessed 19 August 2019.

Laperche, B., and Liu, Zeting. (2013). 'SMEs knowledge-capital formation in innovation networks: a review of literature', Journal of innovation and Entrepreneurship, 21 (2).

Lundström, A., and Zhou, C. (2011). 'Promoting innovation based on social sciences and technologies: the prospect of a social innovation park', Innovation: The European Journal of Social Science Research, 24 (1–2), 133–149.

Ministry of Economic Development and Technology. (2018). 'Akcijski načrt "Slovenija – Dežela inovativnih zagonskih (startup) podjetij"'. <http://www.mgrt.gov.si/fileadmin/mgrt.gov.si/pageuploads/Stasa/AKCIJSKI_ZADN.pdf > accessed 12 July 2019.

MIT Sloan Executive Education. (2014). 'What is really driving academic entrepreneurship?' MIT Sloan Executive innovation@work Blog. <https://www.igi-global.com/dictionary/entrepreneurial-mission-of-an-academic-creative-incubator/47148> accessed 12 May 2019.

Nowotny, H., Scott, P. and Gibbons, M. (2001). Re-Thinking Science: Knowledge and the Public in An Age of Uncertainty. Cambridge: Polity Press.

Peterková, J., and Wozniaková, Z. (2015). 'The Czech innovative enterprise', Journal of Applied Economics Sciences, 10, 243–252.

Philpott, K., Dooley, L., O'Reilly, C., and Lupton, G. (2011). 'The entrepreneurial university: Examining the underlying academic tensions', Technovation, 31 (4), 161–170.

Pustovrh, A., Jaklič, M., Martin, A. S., and Rašković, M. (2017). 'Antecedents and determinants of high-tech SMEs' commercialization enablers: opening the black box of open innovation practices', Economic Research – Ekonomska Istraživanja, 30 (1), 1033–1056.

Savolainen, K. (2013). Nanosafety in Europe 2015–2025: Towards safe and sustainable nanomaterials and nanotechnology innovations. Helsinki: Finnish Institute of Occupational Health.

Serbanica, C. (2011). 'A cause and effect analysis of university – Business cooperation for regional innovation in Romania', Theoretical and Applied Economics XVIII, 10 (563), 29–44.

Shane, S. (2004). Academic entrepreneurship: University spinoffs and wealth creation. Cheltenham: Edward Elgar Publishing.

Siegel, D. S., and Wright M. (2015). 'Academic entrepreneurship: time for a rethink?' British Journal of Management, 26 (4), 582–595.

Tchalakov, I., Mitev, T., and Petrov, V. (2010). 'The academic spin-offs as an engine of economic transition in Eastern Europe. A path-dependent approach', Minerva, 48 (2), 189–217.

The Lund Declaration: Europe must focus on the grand challenges of our time. (2009) Swedish EU Presidency. <https://era.gv.at/object/document/130> Accessed 2 July 2019.

Van Lieshout, M., Enzig, C., Hoffknecht, A., Holtmannspötter, D., Noyons, Ed. and Compano, C. (2008). Converging Applications Enabling the Information Society. Düsseldorf: Zukünfte Technologien Consulting.

Von Schomberg, R. (2012). 'Prospects for technology assessment in a framework of responsible research and innovation', In Dusseldorp, M. and Beecroft R. (eds), Technikfolgen abschätzen lehren, pp. 39–61. Springer VS, Verlag für Sozialwissenschaften.

Westlund, H. (2011). 'Multidimensional entrepreneurship : theoretical considerations and Swedish empirics', Regional Science Policy & Practice, 3 (3), 199–218.

## Data sources

Eurostat. (2019) Science, technology and innovation database. <https://ec.europa.eu/eurostat/web/science-technology-innovation/data/database> accessed 12 September 2019.

# Part 3 Challenges of Governance

Janja Mikulan Kildi

# Democratization and Authoritarianism in the Middle East: Dominant Paradigms and New Perspectives

**Abstract:** This chapter explores the evolution of the dominant scholarship on the politics of the Middle East and North Africa (MENA) region, with the focus on its main strengths and limitations in comprehending post-uprising political trajectories. Despite the popular uprisings of 2010/2011 that generated significant (political) changes, authoritarianism continues to persist across the region. What the previous scholarship largely failed to understand are the nuances of the regimes across the MENA region and, in particular the dynamics - ranging from significant transformations as well as continuities - that eventually affected the uprising's outcomes and post-uprisings political trajectories. Therefore, the main objective of this chapter is to examine theoretical framework and concepts from previous literature on democratization and authoritarianism and point to new perspectives that have to be considered in order to comprehend post-uprising political environments across the region. In two sections, the chapter will provide insights into two main paradigms through which the MENA region and its politics have been explored – 'democratization' and 'post-democratization' paradigms. It will offer a brief look into historical developments of both paradigms, their main features and limitations in explaining post-uprising developments, and how they were shifting. Based on these insights, the final section will offer a discussion on particular concepts related with certain structures, institutions and agency that might aid in comprehending the authoritarian, hybrid and non-authoritarian regimes across the MENA region.

**Keywords:** Authoritarianism, democratization, paradigms, theoretical frameworks, concepts, Middle East and North Africa (MENA)

## Introduction

While authoritarian resilience and democratic deficit are the key features of political orders in the Middle East and North Africa (MENA) countries, they do not make the countries within the region as exceptional as portrayed by public as well as academic discourses or at least not exceptional for the reasons highlighted in the scholarship. In general, the overview of different measures of democracy in 2019 revealed that the map of political regimes across the globe is extremely diverse and dynamic. While by now it became clear that 'the end of history' notion (Fukuyama 1992), which assumed the global dominance of the liberal democracy has not been realized and a third wave of autocratization (while relatively mild so far) is unfolding (Diamond and Plattner 2015; Luhrmann and Lindberg 2019), outlooks on political processes across the globe are different and even contradictory (Adam and Tomšič this volume). In 2010/2011, the MENA region witnessed an incredible mobilization of protestors who challenged incumbent authoritarian regimes across the region. Even though the popular uprisings generated remarkable changes that the region has not seen since the end of the colonial era, autocratic and non-democratic regimes persist. In fact, for almost half of a century (except for specific relatively short periods), 'authoritarian resilience' has been regarded as the most significant feature of politics in the MENA region, especially in comparison to other regions that have been politically transformed during the third wave of democratization in the 1980s and 1990s.

Consequently, the rapid political changes and significant regional implications in 2010/2011 were largely unanticipated. The political trajectories revealed deficient theoretical frameworks and concepts, and the limits of dominant methodological approaches that have been applied in the subfield of comparative politics. As such, the 2010/2011 uprisings not only confronted the existing political orders in most of the countries across the region but also challenged the dominant scholarship, which provided specific frameworks that have not been entirely able to capture complex political dynamic in the field (Bamyeh and Hanafi 2019; Bellin 2012; El Affendi 2017; Elbadawi and Makdisi 2016; Haseeb 2013; Hinnebusch 2018; Makdisi 2017; Saouli 2015; Stepan and Linz 2013; Valbjørn 2015; Valbjørn and Volpi 2014; and many others). More precisely, these scholars are questioning the adequateness of (global) theoretical frameworks, concepts and research methodologies for explaining the political dynamics in the region. In general, there have been two main strands of scholarship tackling the question of politics in the MENA region – the 'democratization' paradigm and the 'post-democratization' paradigm, each of which not only developed entirely different entry points to study politics in the region but also remained analytically separated. While attention will be given to the paradigms that developed in the last three decades (since the 1990s), we will also offer brief but crucial insights into paradigms that developed after the Second World War and served as inspiration for the research frameworks under scrutiny.

Therefore, the main objective of this chapter is to examine the dominant paradigms that are exploring the politics of the MENA region and point to new perspectives that appear meaningful for comprehending post-uprising political trajectories. In order to achieve these goals, the chapter will: a) provide an insight into the evolution of 'democratization' and 'post-democratization' paradigms, their main features and limitations in explaining politics of the MENA region in general, and post-uprising developments in particular; b) offer a discussion on concepts – more precisely, particular structures, institutions and agencies – that might help us comprehend the authoritarian, hybrid and non-authoritarian regimes across the region.

## 'Democratization' paradigm and its roots

The 'democratization' paradigm emerged in the 1990s and explored the regional politics from the perspective of 'democracy' since it has emerged in the context of 'de-orientalization' or 'normalization' of the scholarship on the region and the third wave of democratization. While one strand within this paradigm has been expecting the transition to democracy ('transition' paradigm or 'transitology' in political science), the other strand attempted to explain the democratic deficit. In the context of global anticipations regarding democratic transitions, scholars within the first strand, which has been perceived as the 'democracy-spotting' or almost 'demo-crazy' strand, have been optimistically pointing to specific political developments that arguably imply democratic transitions. Some of these included the crisis of legitimacy of authoritarian regimes (Hudson 1988); the establishment of democratic institutions (parliaments, multi-party systems) and democratic processes (elections) and the emergence of civil society, among others. (Valbjorn and Bank 2010). As underscored by Valbjorn and Bank (2010), this trend was not only related with critiques of MENA scholarship as being some kind of a new form of Orientalism but also to a broader academic debate on globalization as a form of homogenization that will make local knowledge and area studies unnecessary. According to such a line of thinking, this strand of scholarship suggested that the MENA region should be regarded as any other and that studies on MENA politics should be able to draw on insight from and contribute to the general theoretical debates in the fields of political science, (political) sociology, and similar. While this strand of scholarship pointed to some significant political improvements in terms of liberalization, the data revealed that the democracy deficit in the (Arab) Middle East had become even more severe. Therefore, it became apparent that the application of general theoretical models and concepts from the democratization scholarship (e.g., power-sharing, vibrant civil society, frequent elections, economic liberalization, and the emergence of a middle class, etc.) to the MENA context is not sufficient for understanding political dynamics in the region (Valbjorn and Bank 2010).

The second strand of scholarship that emerged in the 1990s has been more realistic and attempted to explain the lack of democratization in the region by focusing on the absence of particular concepts arguably necessary for democracy, such as civil society, a robust middle class, democratic political culture, and democratic interlocutors. In addition, while the MENA region has been largely ignored in general academic and policy debates on democracy promotion, as well as influential cross-regional comparative studies during the 1990s, the notions about Arab/Islamic 'exceptionalism', especially in terms of incompatibility between democracy and Islam, became deeply embedded in scholarship; therefore, they also deserve attention in this overview. Before presenting this strand of scholarship, we will offer a brief insight into pre-existing theories of democratization that influenced the 'democratization' paradigm of the 1990s.

## Predecessors of the 'democratization' paradigm

The dominant structuralist theoretical approaches to democratization in the 1960s and 1970s (Almond and Verba, 1963; Lipset, 1959; Moore, 1966) were widely influenced by the classical modernization theory (CMT), which examines how different societies progress and which variables affect this progress. Essentially, most of the structuralist scholarship on democratization has been influenced by Seymour Martin Lipset's seminal work 'Some Social Requisites of Democracy' (1959) in which he emphasized a correlation (not a causation) between high levels of economic development and democracy. Regardless of the influence of the modernization paradigm in the democratization scholarship, the correlation or causal effect between modernization/economic development and democracy remains highly contested (Chiran, this volume). For example, Cheibub et al. (1996) and Przeworski (2006) determined that authoritarian regimes in growing economies are less likely to experience democratic transition and argued that development makes democratization endure, but it does not make it more likely to emerge. Another critique can be found in the two studies conducted by Acemoglu, Johnson, Robinson and Yared (2008 and 2009; quoted in Makdisi 2017) who argued that over longer periods higher income and democracy are indeed positively correlated; however, there is no evidence of a causal effect. The crucial problematic aspect of modernization theory is its deterministic, unilinear evolutionary and ethnocentric perspective on social change (Hinnebusch 2006). Certain limitations of the modernization paradigm for democratization are also visible when applied to the MENA region. Looking only at economic development does not offer comprehensive explanations for the democratic deficit since the region is economically (and socially) extremely diverse. While some countries have relatively low GDP per capita, data by the World Bank and International Monetary Fund demonstrate that the MENA region is not the most impoverished in the world and that there are many other countries (India, Indonesia) and regions (Sub-Saharan Africa, Asia) where democracy emerged with even lower GDPs. In addition, some of the countries with the highest GDP per capita and/or GDP per capita adjusted for purchasing power parity[1] on regional as well as global levels, such as Qatar, United Arab Emirates and Saudi Arabia (oil-rich monarchies[2]), are the most closed authoritarian regimes, while countries that have comparably much lower incomes, such as Tunisia, Lebanon and Turkey, have much higher democracy scores (World Bank and International Monetary Fund). In addition, the Human Development Index of the United Nations, which also considers health and education (in addition to income), demonstrates that comparatively most of the MENA region falls in the middle level and consequently illustrates that these dimensions might not necessarily explain the deficient democratization. The post-1970s era has seen the theoretical shift in studies of democratization, especially due to several examples of democratization in countries without the presence of structural conditions that have been assumed as prerequisites for democratic processes. The so-called actor or procedural-oriented paradigm assumes that no level of socio-economic or cultural development constitutes a necessary precondition for the initiation of democratic transition (Brumberg 2014). This strand of scholarship has been inspired by the transitions to democracy in the Iberian

---

1   i.e., differences in living costs across countries.

2   However, there are two crucial points to be considered. In the past (before 2000s), the notion of 'oil curse', especially in relation to democracy, received little attention outside the scholarship on the MENA region, and it was not carefully tested with regression analysis (within or beyond the region) (Ross 2001). But since 2001, hundreds of academic studies robustly confirmed that oil has indeed several harmful effects for democratization (Ross 2015). Further, scholarship on the 'resource curse' has also offered significant insights into mechanisms that enable rentier states, particularly those based on oil, to enhance their authoritarian rule (Ross 2015; Elbadawi and Makdisi 2016). However, there are still several aspects that has to be tackled in order to understand the 'resource curse', such as the scope of resource effect, certain aspects related with conditions and mechanism, and solutions for this 'curse' (Ross 2015). The second point is related with the survival of monarchies. The uprisings have implied that there are several additional conditions, that previous literature has largely overlooked, but are enabling monarchies to remain extremely resilient (Yom and Gause III 2012). In addition, such findings can be particularly significant since they not only reveal the dynamics that underlie the success of particular regime type or institutional structure – monarchy – but might also shed led on conditions important for the resilience of autocracies in general.

Peninsula and Latin America during the 1970s and 1980s, and gained confidence with the fall of authoritarian regimes in Eastern and Central Europe, as well as certain political reforms in some MENA countries[3] during the 1990s (Anderson 2006). In contrast to the structuralists (notably many institutionalists and neo-Marxists) that emphasize the limited autonomy of the state's personnel and the extent to which they are constrained by the context (Hay 2002, p. 89), the actor-oriented paradigm assumes that democracy can be constructed through appropriate decisions and policies of major political actors. This paradigm embraces the notion that the optimal scenario for elite-led democratization is:

> a 'combination of a) elite divisions inside an authoritarian regime and b) the formation of an alliance between regime liberals and opposition that is both moderate and popularly credible, in order to marginalize the hard-liners on both sides and incorporate masses. (Hinnebusch 2006, p. 387)

Some scholars of this approach, such as O'Donnell et al. (1986), Linz and Stepan (1996), among others, argue that the relationships between radicals and moderates in the opposition have also helped to determine whether or not transitions occurred and the likelihood of democratic consolidation. The role of agency is also stressed in the social movement theory, which emerged in conjunction with the literature on the third wave of democratization. This theory focuses on the importance of opposition leaders in creating networks, mobilizing resources, and framing their movement's concerns (Goldstone 2010; Lust-Okar 2003; Wiktorowicz 2000;).

For several scholars (Bellin 2018; Mahdavi 2008; Rakner et al. 2007), the agency-oriented approach is reductionist, primarily because a) it reduces the success or failure of democratic transition to psychological features of leaders or the adoption of (incorrect) policies; such an approach largely disregards the historical context and other political and socio-economic structures in place that proved to be essential in comprehending the elite's behaviour in various democratic transitions and consolidations around the globe; b) it primarily focuses on elites. In addition, the focus on leadership is also perceived as being 'atheoretical' since it is driven by individual characteristics (charisma, persuasiveness, social networks, and preferences) that are impossible to predict (Adeny and Wyatt quoted in Bellin 2018). Critiques point out that focus on elites underestimates the role of civil society (Mahdavi 2008) and some of the other actors (such as the

---

3   Which have not led to democracy in most cases but have fast resulted in dynamics such as cancellation of elections (Algeria), outbreak of civil wars (Algeria, Yemen), restrictions of opposition (Tunisia and Jordan) to name just a few (Anderson 2006).

military) as well as the fact that it ignores the importance of interactions between these agents for the success or the failure of democratic transition. While the actor paradigm might offer interesting individual insights about the behaviour of incumbent authoritarian leaders, we cannot overlook the fact that scholars within this paradigm have been too optimistic in interpreting certain (democratic) changes that even proved to reinforce authoritarian political orders. In fact, political dynamics in the MENA region actually broadly challenged their expectations, which 'interprets initial moves towards greater liberty as a kiss of death for autocratic rulers' (Schneider 2009, p. 33).

## Dominant discourses: 'exceptional' region

The MENA region has been overwhelmingly studied through specific lenses (cultural and religious exceptionalism; violence, conflicts and wars, often understood as sectarian; authoritarianism, the applicability of the 'Western' type of democracy), which contributed to largely distorted perceptions of the region. One of the most dominant argument claims that the MENA region is exceptionally resistant to democratization due to its culture/religion (Kedourie 1992). Because the region did not keep pace with other regions in dismantling established orders (Makdisi 2017) and remains the 'last bastion' of authoritarianism in the world, the Islamic and/or Arab 'exceptionalism' arguments are actually deeply embedded and reinforced in the literature on the democratic deficit in the region (Salame quoted in Valbjørn and Bank 2010; Stepan and Robertson 2003). In addition, various cross-regional studies illustrate that a critical mass of democracies exists in every major world region and every major cultural environment except for one – the Arab Middle East. While such observations are important, they might incorrectly imply that the democratic deficit is a result of exclusively cultural or religious attributes. The arguments about the incompatibility of Islam and democracy, have been largely influenced by Huntington's (1996) famous claims about the 'clash of civilizations' and about Islamic doctrine and organization 'not being hospitable' to democracy. However, three main perspectives directly challenge the claim about the 'inhospitality' of Islam to democracy.

First, there is a general agreement that religions (and cultures) cannot be considered as a monolith, uniformed or fixed. It is hard to imagine that such rich social and cultural tradition as Islam, which emerged fourteen centuries ago, would (politically) manifest in only one particular way. It is practically impossible that over 1.6 billion followers, living in every part of the world with different cultural backgrounds and practices, would have precisely the same notions of how to apply the Quaran, Sunnah and other rich traditions, to their daily life and politics. It is also important to note that neither the Quran nor Sharia prescribe or imply a specific form of government or state (Humphreys 2005).

The second perspective is related to the doctrine and certain concepts within Islam. When considering the theology of Islam, both opponents and supporters of the incompatibility paradigm can find arguments for their claims in selective readings of Islamic texts and traditions, and diverse political thoughts of numerous Islamic thinkers. Most of the scholars who rely on the 'exceptionalism' argument point out that one of the most significant obstacles for democratization within Islam is the interpenetration of religion and state or that church/state dualism never emerged and that Islam was the state from the very start (Gellner 1992 quoted in March 2015). However, the political realities across the MENA region are not in accordance with such notions – there are various empirical cases of separation and even domination of politics in the politics-religion relationship (Humphreys 2005; Platteau 2011). Furthermore, while acknowledging that democracy is not something essential in Islam, we can find convincing arguments pointing to certain democratic features of Islam, especially in the second school of thought on Islam and democracy that prevailed in the mid-1980s and has been led by contemporary Islamic thinker Rachid Ghannouchi. They point to concepts such as shura (consultation/consultative council), baia (allegiance), ijma (consensus), al-hurriyya (freedom), al-huqquq al-shar'iyya (legitimate rights) as being entirely consistent with democracy (Ahmad 2011; Jawad 2013).

The third perspective that challenges the notion about the incompatibility of Islam and democracy is related with various public opinion surveys such as World Values Survey and Arab Barometer surveys, public opinion polling conducted by Pew Research Center, as well as national and cross-regional studies based on these surveys. Linz and Stepan (1996) suggest that where democracy is the 'only game in town', meaning that it is highly valued, the possibilities for increased pressure for political liberalization, democratization and consolidation of democracy are higher (Lust 2011; Tessler 2002). While making conclusions about the prospects of democracy exclusively on public opinion surveys has several limitations, they can still offer us some important insights about societal perceptions of democracy and politics in general. In the two most recent waves of Arab Barometer's polling (2016–2017 and 2018–2019), the immediate pursuit of democratic political systems was not listed as a top-three priority, but Arabs still increasingly consider democracy to be the best system of government (Arab Barometer).

## Main critiques of the 'democratization' paradigm

After the entire decade of the 1990s brought no significant improvements in terms of democratization, and expectations about democracy looked like 'waiting for Godot' (Albrecht and Schlumberger 2004), the optimistic strand of the 'democratization' paradigm was faced with several critiques that can be categorized in two main groups (Valbjorn and Bank 2010; Valbjorn 2014).

First is what Valbjorn and Bank (2010) called 'blindness to the actual continuity in the apparent changes'. They have underscored that democracy spotters have not been aware that several nominal regime-led reforms were not a reflection of democratization but a part of a regime-preserving strategy following Lampedusa's famous dictum 'if we want things to stay the same, we will have to change'. As emphasized by Anderson (2006), the focus on the nature of electoral systems, parliaments, parties and other liberal institutions will not only offer an incomplete but even a distorted understanding of politics. As stated by Brumberg (2003), liberalized autocracies have proven far more durable than once imagined. Several MENA experiences demonstrated that the mixture of guided pluralism, controlled elections, and selective repression in various countries (at the time, such countries included Egypt, Jordan, Morocco, Algeria and Kuwait) is not just a 'survival strategy' adopted by the authoritarian regimes, but rather a type of political system that challenges democratization. Diamond (2010) adds that, in such systems, liberalization is not linear but cyclical and adaptive, with regimes ensuring that the opposition remains disadvantaged and disempowered. In practice, when pressure from civil society (including political opposition) increases, the regime loosens its constraints and, allows more civic and political activity. However, if political opposition becomes too effective and seems able to endanger the regime, then the incumbent returns to more severe mechanisms (rigging elections, shrinking political space, and arresting suspects). Diamond uses the metaphor of lungs to describe such dynamics: 'the electoral arena in these states is something like a huge pair of political lungs, breathing in and expanding, but then inevitably exhaling and contracting when limits are reached' (Diamond 2010, p. 99).

In addition to looking at elections, democratic institutions and civil society, the continuity behind apparent democratic reforms can also be explored through the lens of foreign interventions and democracy promotion (Yom and Al-Momani 2008; Yom 2015).

The second group of critiques is related with a tendency of democracy spotters to perceive the absence of regime change as a 'standstill' (Valbjorn and Bank 2010) and consequently fail to observe some critical political changes, such as a) transformations within the regimes, such as the shift from populist to post-populist authoritarianism (Hinnebusch 2006); b) the changes of various relationships (regime vs opposition, especially Islamist parties (Schwedler 2006); relationships within the opposition (emerging cooperation between the secular Left and Islamists); and changes in the larger society, like the emergence of alternative orders.

In sum, due to such dynamics, the 'democratization' paradigm became 'doubly isolated' (Valbjorn and Bank 2010). First, its adherents have been isolated from the 'desired' discipline from two perspectives: a) due to lack of successful democratic transitions, the MENA region has been largely neglected in comparative studies on democracy and democratization; b) at the same time, certain theoretical models and concepts from the general democratization theories proved to have negative implications for democratization (Valbjorn and Bank 2010): for example, the concept of power-sharing derived from Spanish and Latin American transitions (Anderson 1999); the emergence of a vibrant civil society (Schlumberger 2000; Wiktorowicz 2000); the frequent elections that were primarily 'voting for wasta[4]' (Lust 2009; Schwedler and Chomiak 2006); economic liberalization that did not encourage the middle class to demand democracy but contributed to the rise of crony capitalists who were not interested in broader political liberalization (Hinnebusch 2006).

Second, this strand also became isolated from their 'own' region since the 'democracy' perspective did not provide appropriate tools to understand regional politics (Valbjorn and Bank 2010). By focusing on democratic dimensions (especially democratic institutions and processes, and merits of democratic or authoritarian regimes for economic development) scholars have been mostly searching 'where the light shines' (Anderson 2006) and have not been able to develop frameworks for capturing significant political actors that challenged the regimes (like transnational Islamist movements and ethnic communities) and dynamics that affected the politics (Saouli 2015).

---

4   *Wasta* is a widespread concept in the Arab region that refers to nepotism or having different kinds of connections (family, tribal or other social affiliations). The literal meaning is 'intercession' or 'intermediation'. *Wasta* dynamics are crucial in determining who will be the recipient of allocated rents and who will be given greater commercial, career and institutional advantages, especially by bypassing official channels (Gray 2019).

## Second 'post-democratization' paradigm and its roots

Due to several limitations of the 'democratization' paradigm that we explored in the previous chapter, scholarship in the 2000s witnessed a paradigm shift from 'democratization' to 'post-democratization' that focused on 'authoritarian resilience'. Scholars within this strand (Anderson 2006; Bellin 2012; 2018; Heydemann 2002; Kramer 2001; Mitchell 2003; Schlumberger 2007) directed their attention to how political rule in Arab countries is 'effectuated, organized and executed to generate stability and regime survival' (Saouli 2015). The paradigm provided important insight into the nuances of authoritarianism and mechanisms that regimes employed to remain in power (Valbjorn 2014).

In order to understand the roots of the second 'post-authoritarian' paradigm, we need to go to the origins of the concept of 'authoritarianism', which started to appear in the 1960s, when the 'totalitarianism' paradigm (1930s/1940s–1960s) started to face increasing critiques due to its inherent statism, fuzzy conceptual foundations and scarcity of empirical cases (Gerschewski 2013).

The rise of authoritarianism as a distinct regime type began with O'Donnell's study on 'bureaucratic authoritarianism' (1973), which argued that, because Argentina and Brazil reached limits in their import substitution strategy, bureaucratic elites from the military and business formed a coup coalition and established an authoritarian regime because they were frustrated with the political and economic crisis. Although these explanations were not viable in other Latin American countries or beyond, O'Donnell emphasized the importance of socio-economic dimension for the emergence and the survival of authoritarian regimes, which had largely been overlooked until then (Gerschewski 2013). In comparison to the totalitarian paradigm, this strand of scholarship offered more tailored or regional explanations for new phenomena (Gerschewski 2013).

With regard to MENA scholarship, this period can be labelled as the first post-democratization era (1960–1980), which has been the response to the gap between the democracy-spotting scholarship of the 1950s (first demo-crazy studies on the MENA region) and realities on the ground. Just like in the 1990s, democracy spotters of the 1950s were looking for any kind of signs that would transform MENA regimes into models of Western liberal democracies. In his seminal study, Daniel Lerner (1958) assumed (similarly as Fukuyama three decades later) that the same basic participation model would reappear in all modernizing societies across the globe regardless of racial, ethnic and religious differences (Valbjorn and Bank 2010). However, during the 1960s, such optimistic expectations were replaced with disappointments since political realities across the region involved civil wars, recurrent military coups, the abolition of (nominal) multiparty systems in favour of absolute monarchy/single-party states, and similar. (Valbjørn and Bank 2010). Scepticism about establishing democracy in the MENA region has been reflected in declining scholarly interest in democracy and instead re-focusing its attention to the institution-building/ political order (Huntington 1968) and external exploitation of authoritarian regimes by the Western powers in the global capitalist system (neo-Marxist approach focused on dependency and world system(s) theory).

There was more than a decade (between the late 1980s and the 2000s) with very few systematic studies on authoritarianism across the world. The third wave of autocracy research started in the late 1990s when a number of studies, many of them using newly assembled large-N datasets and employing statistical techniques (especially regression), started to be published. In particular, Barbara Geddes' (1999) path-breaking study of the durability of different types of authoritarian rule,[5] inspired a new generation of comparativists to study autocracies (Gerschewski 2013; Kollner and Kailitz 2013). A particular focus of this strand of scholarship was given to co-optation and strategic repression. From the perspective of co-optation, scholars have emphasized the systematic use and stabilizing effects of democratic institutions and processes for the survival of authoritarian regimes (neo-institutionalist approaches) (e.g., Gandhi and Przeworski 2006; 2007). Elections have been seen as an essential tool for co-optation, while political parties seemed to have a stabilizing effect on autocratic rule, as they were able to settle and mediate intra-elite conflicts, alleviate moral hazard problems and maintain intra-elite cohesion (Gerschewski 2013).

In contrast to the repressive abuse in totalitarian regimes, several scholars emphasized that even though repression is one of the main mechanisms for authoritarian survival, it cannot account for the longevity of regimes alone. Acemoglu and Robinson (2006) pointed out that what is particularly decisive in the durability of autocracies are strategic or 'optimal' degrees of repression. For example, Lewitsky and Way (2002) made a useful distinction between 'high'[6] and 'low intensity'[7] repression according to the targeted people or institution and the form of violence used. Even when the security apparatus is able to repress thousands of people violently, such repression is usually very costly since it may

---

5  Based on earlier qualitative literature, Geddes (1999) distinguished military, one-party and personalist dictatorships.

6  Visible acts that are targeted either at well-known individuals like opposition leaders, at a larger number of people, or at major oppositional organizations.

7  Aimed at groups of minor importance, is less visible, and often takes more subtle forms.

hinder the institutional integrity of the security forces, domestic legitimacy and international support, especially in case of intense mobilizations (Bellin 2004).

Together with the third wave of general autocracy studies, MENA scholarship also saw the emergence of a 'renaissance' in the study of authoritarianism or second 'post-democratization' paradigm. Kramer (2001) and Mitchell (2003) argued that the focus of the 'democratization' paradigm on so-called 'Western categories' and 'Western discourse' and the neglect of certain aspects, especially the roles and impacts of Islamism, distorted understanding of the political dynamics in the region. In her prominent piece, Bellin (2004, p. 148) argued that 'the solution to the puzzle of Middle Eastern and North African exceptionalism lies less in absent prerequisites of democratization and more in present conditions that foster robust authoritarianism, specifically a robust coercive apparatus in these states'. In general, the scholars within the 'post-democratization' paradigm conducted highly regime-oriented studies and mostly perceived society as largely de-politicized. While the paradigm with the main focus on authoritarian resilience seemed like an adequate research approach, the 2010/2011 uprisings have challenged some of their assumptions.

In the Tab. 1, we present the overview of scholarship on democratization and autocracy, in general, and scholarship on the politics in MENA, in particular.

## New perspectives on MENA politics after the 2010/2011 uprisings

While the post-democratization paradigm 'may have contributed to bringing politics back into the analysis, its relatively narrow focus on regime resilience overlooked different forms of resistance against domination of authoritarian regimes' (Saouli 2015; Schlumberger 2007). Some critiques of the 'post-democratization' paradigm appeared prior to the uprisings. For example, Anderson (2006, p. 10) believed that the study of authoritarianism is 'little more than the obverse of the 'inevitability of democracy', inflected with 'pessimism' since it has largely directed attention to the same kinds of research subjects (e.g., the strength or weakness of civil society, or regime manipulation of interest groups). She also believed that there has been a tautological character to this argument, which she does not consider surprising since 'authoritarianism' is 'little more than a residual category in most political science, encompassing all the otherwise very varied nondemocratic regimes that have existed throughout history' (Anderson 2006, p. 201). In parallel with the post-democratization paradigm in the 2000s, another, otherwise less influential strand of scholarship on the

**Tab. 1:** Overview of scholarship on politics in the MENA region and beyond until 2010/2011 uprisings

|  | DEMOCRATIZATION PARADIGM | 'POST-DEMOCRATIZATION' PARADIGM |
|---|---|---|
| **1930s/1940s** |  | 'Totalitarianism' paradigm |
| **1950s** | - Heyday of liberal modernization theory: <br> - First 'democracy spotting' or 'demo-crazy' era of scholarship on the MENA region |  |
| **1960s–1970s** | 1 **The modernization strand**, with dominant structuralist theoretical approaches: focus on 'social requisites' of democracy and their absence; <br> 2 **Actor-oriented strand** with focus on agency or procedural oriented approaches |  |
| 1970s–1980s |  | - Juan L. Linz's seminal attempts to distinguish authoritarian rule from totalitarian and democratic rule (and also personalist or 'sultanistic' rule); <br> - The emergence of the 'authoritarianism' concept (O'Donnell 1973); <br> - Focus on socio-economic conditions and informal politics + regional specifics (neo-patrimonial rule and 'social contract' in Arab world) |
|  |  | **MENA region (early 1960s–1987)** <br> - First post-democratization paradigm as a response to 'demo-crazy' era of the 1950s: focus on institution building (political order) and external exploitation of the regimes in the MENA region by Western powers |

**Tab. 1:**   Continued

|  | DEMOCRATIZATION PARADIGM | 'POST-DEMOCRATIZATION' PARADIGM |
|---|---|---|
| 1990s | **Two opposing strands in the MENA scholarship:**<br>– Second 'democracy spotting' or 'demo-cracy' era in the context of third wave of democratization and 'de-orientalization' or 'normalization' strand: searching for the signs of democratic transition in the region;<br>– Revival of 'exceptionalism' claims | **Stagnation of research on autocracies**<br>Very few systematic studies applying or building on existing theoretical frameworks and concepts |
| 2000s |  | – 'Renaissance' of autocracy studies: focus on co-optation and 'strategic' repression<br>– **Second post-democratization paradigm (MENA):** the mixture of guided pluralism, controlled elections, and selective repression |
| 2011 – |  | New paradigm which links 'democratization' and 'post-democratization' paradigm |

MENA region suggested that there is a need for the expansion of scholars' under-standing of what constitute the relevant political agencies, what counts as poli-tics, and where politics takes place (Valbjorn and Bank 2010). Such perspectives influenced studies that are attempting to move beyond 'democratization' and 'authoritarianism' paradigms in favour of looking above, below and beyond the level of the regime (Valbjorn and Bank 2010). As stressed by Valbjorn (2014) this more society-centred focus was not associated with any expectations of significant regional uprisings but focused on growing gaps in the state-society relationship.

The 2010/2011 uprisings, their outcomes and post-uprising trajectories implied that the political dynamics in the MENA region cannot be adequately explained by isolating one explanatory paradigm from the other. As stressed by Yom (2015), the uprisings were followed by an academic publishing boom. Within the first five years 'more than two hundred English-language academic books and over a thousand journal articles in all languages concerning the uprisings were published, accompanied by large edited volumes with country-specific chapters' (Yom 2015). Initially, the optimistic strand assumed that the MENA region is in a transition to democracy.[8] However, very soon, and especially in the first five years after the uprising, the 'post-paradigm' assumptions were confirmed since the majority of regimes where uprisings unfolded remained autocratic or even enhanced their authoritarian rule. Based on the dialogue between the dominant scholarship on the MENA region and empirical post-uprisings development, we can conclude that both paradigms – 'democratization' and 'post-democratization' – have failed to offer a comprehensive framework that would enable understanding politics as well as nuances of regimes across the region – authoritarian (hard-line and moderate), hybrid and non-authoritarian. Adopting a particular paradigm narrows the lenses through which we consider political trajectories and prevents us from developing such a comprehensive framework.

Limitations of existing scholarship are mainly related with the following dimensions: a) struggles of scholarship to escape from the 'democratization' vs 'authoritarian resilience' approach or democracy-authoritarianism dichotomy (Valbjorn 2015); b) immediately after the uprisings, most scholars sought to understand what had been missed by the literature on authoritarian resilience and consequently remained at the level of agency (anti-regime mobilization; the strategies of elites), while those examining the structural context have focused on how grievances have been generated rather than their impact on post-uprising developments (Hinnebusch 2015); c) while understanding the causes of uprisings is significant, it does not tackle the fact that these significant processes did not evolve in in a vacuum and that their nuanced outcomes are the result of constellations of various dimensions rather than only particular aspects (Hinnebusch 2016; 2018; Yom 2015); d) there is a need to move beyond mostly descriptive reviews of events and towards more explanatory approaches (Yom 2015).

In order to tackle the first critique, it is crucial to acknowledge the nuances of regimes across the region that have been neglected to a certain extent. While the 'post-authoritarian' paradigm offered important insights into different mechanisms employed by different forms of authoritarian regimes, they have largely neglected various regimes that are hybrid and non-authoritarian. In

---

8   Despite some optimistic assumptions, Hudson (2011) stated that this time the position is still more realistic and not so demo-crazy as it once seemed.

order to obtain a comprehensive understanding of the region, these regimes also must be included in the comparative.

Moreover, the uprisings and especially the post-uprising trajectories have confirmed the importance of acknowledging the interplay between structure and agency – but how to frame such an interplay? The limitations of both structuralism and intentionalism (privileging agential factors) have suggested moving beyond these extremes to some middle ground. From the late 1970s onwards, a new generation of social theorists, such as Anthony Giddens, Jeffrey Alexander, Roy Bhaskar, Margaret Archer, Pierre Bourdieu and Piotr Sztompka, effectively argued that structuralism and intentionalism had failed to deal with the relationship between structure and agency by simply reducing one to the other. What was required was a return to the most basic of ontological principles, those concerning the actor and the context in which he/she finds himself/herself. It is argued that what is required is a mode of analysis (and corresponding social ontology) capable of reconciling structural and agential factors within a single explanation; an account which is neither structuralist nor intentionalist yet an account which does not merely vacillate between these two perspectives (Hay 2002, p. 113). Despite the common ontological core, however, the precise view of the relationship between structure and agency and the implications one might draw from it for political analysis of power and change vary considerably on the two approaches, namely Giddens' structuration theory and the critical realism of Bhaskar and Archer. Through a critical engagement with these highly influential positions, Bob Jessop (2008, p. 117) establishes a point of departure for the strategic-relational approach (SRA). Within the SRA, the distinction between structure and agency is a purely analytical one. According to this approach, neither agents nor structures are real, since neither has an existence in isolation from each other – their existence is relational (structure and agency are mutually constitutive) and dialectical (their interaction is not reducible to the sum of structural and agential factors treated separately). While it may be useful analytically to differentiate between structure and agential factors, this analytical distinction must not be solidified into a rigid ontological dualism (Jessop 2008; Hay 2002; Golob and Makarovič 2019; Golob and Makarovič, this volume).

The third and last point, closely connected to the second one, is related to the fact that while comparative scholarship - both general, as well as specializing on the MENA region – is increasingly acknowledging that political outcomes are the result of the complex, relational and dialectical existence of structures, institutions and agency, so far, we have not been able to identify any research that would explore the importance of such configurations in-depth, especially in terms of 'measuring' the significance of configurations under scrutiny. Analysis of such constellations might also contribute to making a shift from mainly descriptive to more explanatory studies of the region.

## Tackling critiques and providing new insights

The 2010/2011 uprisings have confirmed that the fall of the incumbent ruler does not equal regime change and/or initiate the transition to democracy. While political dynamics cannot be fully comprehended without looking above, below and beyond the level of the regime, we believe that such an approach needs to avoid neglecting the mechanisms that regime employs to obtain or even upgrade their power. At the same time, special attention needs to be given to the society level, especially because uprisings and several other dynamics in the region proved that societies across the region can no longer be perceived as politically apathetic and de-politicized, and that, consequently, the mechanisms employed by the regimes might have different implications for society than in the past. Therefore, in relation to the MENA region and post-uprising trajectories, the following structures, institutions and agency might have particular explanatory power:

- Structural factors tackled through the concepts of 'state capacity' (stateness and government effectiveness); autonomy of civil society; 'rentier' state (political economy); and transnational dynamics (with particular focus on regional structure of power);
- Institutional factors: the dynamics within the military;
- Agential factors tackled from the perspective of a regime and their mechanisms to remain in power (with special attention on consensus building, political terror and securitization).

According to several scholars (Malmvig et al. 2016; Saouli 2015), one of the most significant trajectories in the post-uprisings period is the erosion of state capacities. Throughout the decades, notions about the 'strength' of the state have been fluctuating. In the first decades after the European decolonization, several scholars (Halpern 1963; Hudson 1977) perceived MENA states as 'artificial' and 'weak' since their political boundaries have been imposed by the colonial powers and did not reflect pre-existing communities (Malmvig et al. 2016). Such perceptions were challenged from the mid-1980s and during the 1990s when scholars (Anderson 1987; Harik 1990; Luciani 1990) started to acknowledge that states had become relatively stabilized and consolidated political units. Harik (1990) even argued that states had not been as 'artificial' as assumed but had been based on pre-colonial political, economic and cultural power centres. Anderson

(1987), inspired by the work of Skocpol's (1985) famous notion 'bringing the state back in', suggested that studies on the MENA region should become more 'state-centred'.

During the same period, the theory of the 'rentier state', which enables regimes to remain in power through different mechanisms (welfare, co-optation, repression) and will be explored further on, have also emerged. In sum, the majority of scholars in this period emphasized the importance of the state and portrayed societies as being repressed and weakened by the regime. From the 2000s onward, the assumptions regarding state capacities have shifted yet again due to several domestic and transnational socio-economic and political dynamics. The social contract in several rentier states has been challenged by population growth, volatile commodity markets and the global pressures of political liberalization. While certain states (such as Algeria) have remained relatively isolated from regional and international interventions, this has not been the case for several other countries such as Palestine, Lebanon, Yemen and Iraq that have been highly penetrated (Furtig 2016). These dynamics have been strengthened during and after the 2010/2011 uprisings when the erosion of state capacities and power vacuums challenged the monopoly of the state and allowed domestic and transnational non-state actors to take control over large parts of the state's territory. In some countries (Syria, Yemen, Libya), such trajectories led to civil wars or substantial turmoil (Iraq, parts of Egypt) (Gause III 2015; Malmvig 2016; Saouli 2015). Therefore the framework for understanding current dynamics in the region cannot disregard such transformations of state capacities both in terms of their state-ness and government effectiveness.

Further, in terms of structure, the uprisings have demonstrated the importance of 'bringing political economy back' (Medani quoted in Haddad and Schwedler 2013) as a key framework of analysis with particular focus on the 'rentier state'. In the MENA region, there are eleven so-called 'rentier states' that depend heavily on rents and can consequently afford various mechanisms that enable the regimes to remain in power. First, rentier states do not need to tax (heavily), and their primary function is to support society through the distribution or allocation of oil rents received from the rest of the world. In contrast to production states, the rentier state is (financially) independent of society and autonomous and does not seek legitimacy via democratic institutions like parliaments (Luciani 2009; Ramsay 2011). Consequently, regimes in such states become alienated from their societies and are not accountable to their citizens (Ross 2015; Elbadawi and Makdisi 2016). In such states, the well-known political demand 'no taxation without representation' is reversed into the political reality of 'no representation without taxation' (Luciani 2009; Ross 2015). There are various mechanisms through which (more) oil revenues lead to more autocratic regimes and deserve special attention: a) maintaining existing social contracts with the population by preserving the welfare state (direct cash transfers, generous pension programmes) and substantial public sector employment (Elbadawi and Makdisi 2016); b) political repression; c) investments in building broad socio-political (elite) coalitions with possible opposition.

In addition to looking at the regime level, there has to be a deeper understanding of dynamics within civil society, including political parties. The overall image, a relatively distorted one, is that civil society in MENA is relatively weak and depoliticized due to various cultural and political dynamics. However, various scholars such as Altan-Olcay and Icduygu (2012), Hardig (2015) and Yom (2015) suggest a re-conceptualization of the civil society that will enable us to: a) understand that the boundaries between state and civil society are not completely determined, especially because civil society actors across the region are often co-opted by the regimes and used as a tool for social control over the public sphere. (Wiktorowicz 2000; Al-Sayyid Sa'id quoted in Altan-Olcay and Icduygu 2012); b) understand the significance of the 'informal' part of the civil society (Yom 2015).

Furthermore, in the context of civil society and especially political party opposition, particular attention should be given to Islamic movements and parties. While religion has not had a significant impact on democratic transitions and transformations during the third wave of democratization, this cannot be said for the social and political transformations across the MENA region that witnessed the emergence and development of strong Islamic-oriented parties. Therefore, it is first crucial to understand the broader relationship between religion/Islam and the state/government; and second, it is important to examine the role of Islamic movements and parties in the transformations of authoritarianism.

The final structural condition is related to the transnational level. While several studies have focused on the impact of international factors on the uprisings and post-uprising trajectories, very few have examined the impact of newly emerging regional political order that had also been marginalized prior to the uprisings (Bank and Valbjorn 2010). Therefore, we believe that the implications of transnational factors should be tackled from different perspectives than has been done thus far, primarily due to the declining role of more traditional powers in the region (especially the U.S.) and the emerging influence of regional powers (Saudi Arabia, Iran, Turkey). However, several scholars (Byman 2013; Hinnebusch 2018) suggest that regime outcomes after the 2010/2011 uprisings could be best explained by looking at the competitive intervention by several rival powers with opposing projects (Hinnebusch 2018) and examining shifting patterns in the distribution of power (Kamrava 2018). As stated by Kamrava (2018), the regional order in the Middle East has consistently demonstrated a significant inclination toward tension and crisis, and uprisings have added new elements and additional layers of complexity to existing patterns of power within and between different regional powers (both status quo and counter-hegemonic states).

In terms of institutions, special attention should be given to the military. While trajectories of the uprisings have been affected by various dimensions, military responses to the events have been the most reliable predictor of the initial outcomes – in Tunisia and Egypt, when militaries decided not to back the incumbents, the regimes collapsed; in Bahrain and Syria when militaries remained loyal to the regime, the regime managed to survive; in Libya and Yemen when the militaries fractured, the outcome was mostly determined by other dimensions. A coercive apparatus, especially the military, is one of the most powerful domestic institutions and supportive pillars of authoritarian regimes across the MENA region. Nevertheless, the scholarly attention on the military's political role has been severely limited (Albrecht and Bishara 2011; Bellin 2004; 2012; Gause III 2011; Taylor 2014). Similarly, the scholarship on the uprisings has mostly ignored the impacts of militaries' (non)interventions on the uprisings' outcomes and post-uprisings trajectories and notably failed to understand the nuanced and puzzling behaviours of the military. Such diverse responses require further investigation of conditions that underlie particular military behaviour (coup-proofing strategies, political and economic interests of the military, legitimacy of the military, transnational support, etc.) and their political implications.

In terms of agency, looking both at the level of the regime and the level of society might be particularly insightful. As already emphasized, there is a dominant notion in MENA scholarship that authoritarian resilience is the outcome of different mechanisms employed by the regime, especially repression and co-optation. While repression/political terror is one of the defining features of authoritarian regimes (consequently, the danger of tautology is present), it is significant to explore this dimension more in-depth, especially because different types of repression may condition the polarization among (opposition) political groups (Nugent 2019). Nugent, who draws her research on psychological theories of social identity, concluded that targeted repression prime increases in-group identification and affective and preference polarization between groups, while a widespread prime decreases these same measurements and increases identification with the outgroup (Nugent 2019). While the scholarship on co-optation has emphasized the effects of democratic institutions and processes for the survival of authoritarian regimes, the perspective that might also help us understand not only the authoritarian stability but also non-authoritarian regimes across the MENA region is the ability of political elites to establish a broad consensus on reform with other actors in society, especially by establishing consensus on political (and economic) goals of transformation, depolarizing cleavage-based conflicts and, excluding or co-opting anti-democratic actors. Democratization is largely dependent on the consensus-building skills of the new regime, especially if democratization is perceived as partially a normative process of social construction' that turns on 'persuasion, deliberation and generation of consent' (Whitehead 2002). As underscored by several prominent scholars (Karl 1990; Przeworski and Limongi 1997, etc.) who advocated the importance of political agency for democratic transitions, 'democratization' is a mechanism for institutionalizing uncertainty by subjecting all interests to competition and that incumbents need to persuade different constituencies (especially those with veto powers, like the military and business community, as well as society at large) to follow the rules of this competition featured by both gains and losses (Bellin 2018).

In contrast to consensus building, regimes across the region have largely adopted the language of securitization in the post-uprising period; therefore, how this mechanism is affecting politics should be explored. In the context of securitization, regimes frame their challenges as existential 'threats' (usually against the survival of the nation-state) and open the sphere to extraordinary means to tackle these challenges. This means that securitization takes the challenges from the sphere of normal politics where challenges would be addressed with dialogue, negotiations and compromise to the sphere of security (Buzan et al. 1998; Malmvig et al. 2016).

## Conclusion

This chapter attempted to illustrate that while 'democratization' and 'post-democratization' paradigms offer several important insights about the politics of MENA region, they both fail to offer a comprehensive framework that would enable us to understand complex political dynamics in the region. The empirical developments in the field, more precisely the post-uprisings trajectories have once again implied that a new shift in paradigms might be necessary in order to comprehend political dynamics in the MENA region. We suggest that the construction of comprehensive framework must be based on extracting and synthesizing the most essential aspects from both paradigms, and offer new insights or what might be described, in the terminology adopted by Bill (1996),[9] as 'new wine in old bottles'. While a rich body of scholarship is accumulating on the post-uprising trajectories and convincingly explaining divergent dynamics and political outcomes[10] in a particular case study or from a particular perspective, very few (if any) succeed in providing a more comprehensive framework that would be able to account for the complex dynamics in each national context but at the same time also not abandon the aim of developing (at least) middle-range theory that might be applicable for the region as a whole.

In order to pursue this objective, we must first go beyond the competition between single explanatory concepts and integrate them into a more coherent set of propositions and test which combinations between them produce specific types of regime in the region – authoritarian, hybrid and non-authoritarian. The social realities are highly complex, and there are many pathways of getting to the authoritarian resilience or non-authoritarian regimes. While the need for such an integrative approach has been repeatedly addressed by various scholars, it has not been followed by many in practice (Schneider 2009). While the chapter has suggested some concrete concepts to be included in the analytical framework – state capacity, rentier state, regional structure of power, dynamics within the military, mechanisms of regimes to remain in power (consensus building, political terror and securitization), the autonomy of civil society and role of the religion – it is crucial that these concepts and especially their configurations be explored within appropriate methodological frameworks. In contrast to dominant research paradigms related to the MENA region, we suggest that the research problems and concepts under scrutiny can be best approached by applying Qualitative Comparative Analysis (QCA) that enables the researchers to tackle and unravel complex causation understood as causation that is conjunctural (one single condition unfolds its impact on the outcome to be explained only when combined with one or more other condition(s); equifinal (allowing for the possibility that

---

9    James Bill (1996) has been very critical of the (American) study of MENA politics and believed that scholars on the region had been wandering in circles and 'producing old wine in new bottles' without making progress in understanding the region.

10   Special issues published by the Middle East Critique (2010) journal suggested four different broad 'post-democratization' avenues (authoritarian learning and extra-regional di/convergence; non-democratic legitimacy; technical modernization without democratization and regional-domestic entanglements), while the British Journal of Middle Eastern Studies also published a special issue (2015) elaborating on the 'scientific contribution that the examination of continuity and change before and after the uprisings can make to our understanding of politics in the region' (Rivetti 2015).

different conjunctions, paths or causal 'recipes' can produce the same (political) outcome), and asymmetric (specific conjunctions producing fragility of authoritarian regimes are not necessarily just the absence of the factors producing authoritarian resilience (Ragin 2008; Schneider 2009; Wagemann and Schneider 2010)).

# References

Acemoglu, D. and Robinson, J. A. (2006). Economic Origins of Democracy and Dictatorship. New York: Cambridge University Press.

Acemoglu, D., S. Johnson, J. Robinson, and P. Yared (2008). 'Income and Democracy', American Economic Review, 98 (3), DOI: 10.1257/aer.98.3.808808-42.

Acemoglu, D., S. Johnson, J. Robinson, and P. Yared (2009). 'Reevaluating the Modernization Hypothesis', Journal of Monetary Economics, 56 (8), DOI:10.1016/j.jmoneco.2009.10.002.

Ahmad, I. (2011). 'Democracy and Islam', Philosophy and Social Criticism 37, 459–70.

Albrecht, H. (2015). 'Does Coup-Proofing Work? Political–Military Relations in Authoritarian Regimes amid the Arab Uprisings', Mediterranean Politics, 20 (1) 36–54, doi: 10.1080/13629395.2014.932537.

Albrecht, H. and Schlumberger, O. (2004). "Waiting for Godot': Regime Change without Democratization in the Middle East', International Political Science Review 25, 371–392.

Albrecht, H. and Bishara, D. (2011). 'Back on Horseback: The Military and Political Transformation in Egypt', Middle East Law and Governance 3(3): 13–23.

Almond, G. A. and Verba, S. (1963). The Civic Culture: Political Attitudes and Democracy in Five Nations. Princeton, NJ: Princeton University Press.

Altan-Olcay, O. and Icduygu, A. (2012). 'Mapping Civil Society in the Middle East: The Cases of Egypt, Lebanon and Turkey', British Journal of Middle Eastern Studies 39, 157–179.

Anderson, L. (1987). 'The state in the Middle East and North Africa.' Comparative Politics 20 (1), 1–18.

Anderson, L. (1991). 'Absolutism and the Resilience of Monarchy in the Middle East', Political Science Quarterly 106 (1), 1–15.

Anderson, L. (2006). 'Searching where the Light Shines: Studying Democratization in the Middle East, Annual Review of Political Science, 9 (1), 189–214.

Arab Barometer. (2019). https://www.arabbarometer.org/topics/governance/

Bamyeh, M. and Hanafi, S. (2015). 'Introduction to the special issue on Arab uprisings. International Sociology', 30(4), 343–347. https://doi.org/10.1177/0268580915584500.

Bellin, E. (2004). 'The Robustness of Authoritariansim in the Middle East: Exceptionalism in Comparative Perspective', Comparative Politics, 36, 139–157.

Bellin, E. (2012). 'Reconsidering the Robustness of Authoritarianism in the Middle East: Lessons from the Arab Spring', Comparative Politics 44, 127–149. doi:10.5129/001041512798838021.

Bellin, E. (2018). 'The Puzzle of Democratic Divergence in the Arab World: Theory Confronts Experience in Egypt and Tunisia', Political Science Quarterly, 133 (3), 435–474. DOI: 10.1002/polq.12803.

Bill, J. A. (1996). 'The Study of Middle East Politics, 1946–1996: A Stocktaking', Middle East Journal 50 (4), pp. 501, 507.

Boix, C. and Stokes, S. (2003). 'Endogenous Democratization', World Politics, 55 (4): 517–549.

Börzel, T. A., Risse, T. and Dandashly, A. (2015). 'The EU, External Actors, and the Arabellions: Much Ado About (Almost) Nothing', Journal of European Integration 37, 135–153.

Brumberg, D. (2003). 'Liberalization versus Democracy – Understanding Arab Political Reform', Carnegie Papers, 37.

Brumberg, D. (2014). 'Reconsidering Theories of Transition'. In Marc Lynch (ed.), The Arab Uprisings Explained: New Contentious Politics in The Middle East. New York: Columbia University Press.

Bromley, S. (1997). 'Middle East Exceptionalism: Myth or Reality'. In: David Potter et al. (eds) Democratization. Cambridge: Polity Press.

Bush, S. S. (2015). The Taming of Democracy Assistance Why Democracy Promotion Does Not Confront Dictators. Cambridge: Cambridge University Press.

Buzan, B., Wæver, O. and de Wilde, J. (1998). Security. A New Framework for Analysis, Boulder: Lynne Rienner Publishers.

Byman, D. (2013). 'Explaining the Western Response to the Arab Spring', Journal of Strategic Studies, 36 (2): 289–320, DOI: 10.1080/01402390.2013.773891.

Carothers, T. (2002). 'The End of the Transition Paradigm', Journal of Democracy, 1 (2002), 5–21.

Cheibub, J. A., Przeworski, A., Limongi Neto, F. P., and Alvarez, M. M. (1996). 'What Makes Democracies Endure?' Journal of Democracy 7, 39–55.

Diamond, L. (2010). 'Why Are There No Arab Democracies', Journal of Democracy, 21 (1), 93–104.

Diamond, L. and Plattner, M. (2015). Democracy in Decline. Baltimore: Johns Hopkins University Press.

Durac, V. and Cavatorta, F. (2009). 'Strengthening Authoritarian Rule through Democracy Promotion? Examining the Paradox of the US and EU Security Strategies. The Case of Bin Ali's Tunisia', British Journal of Middle Eastern Studies, 36 (1), 3–19, DOI: 10.1080/13530190902749523.

Elbadawi, I. and Makdisi, S. (2016). Democratic Transitions in the Arab World. Cambridge: Cambridge University Press.

El Affendi, A. (2017). 'Overcoming Induced Insecurities: Stabilising Arab Democracies after the Spring' in I. Elbadawi and S. Makdisi (eds.), Democratic Transitions in the Arab World. Cambridge: Cambridge University Press.

Epstein, D., Bates, R., Goldstone, J., Kristensen, I. and O'Halloran, S. (2006). 'Democratic Transitions', American Journal of Political Science 50 (3): 551–569.

Fukuyama, F. (1992). The End of History and the Last Man. New York: Avon Book.

Gause, G. F. (2000). 'The Persistence of Monarchy in the Arabian Peninsula: A Comparative Analysis'. In Kostiner, J. (ed.), Middle East Monarchies, pp. 167–86. London: Westview).

Gause, G. F. (2011). 'Why Middle East Studies Missed the Arab Spring', Foreign Affairs 90 (4): 81–90.

Gandhi, J. and Przeworski. A. (2006). 'Cooperation, Cooptation, and Rebellion under Dictatorships', Economics and Politics 18 (1): 1–26.

Gandhi, J. and Przeworski. A. (2007). 'Authoritarian Institutions and the Survival of Autocrats', Comparative Political Studies 40 (11): 1279–1301.

Geddes, B. (1999). 'What Do We Know about Democratization after Twenty Years?', Annual Review of Political Science 2 (1999): 115–144.

Gerschewski, J. (2013). 'The Three Pillars of Stability: Legitimation, Repression, and Co-Optation in Autocratic Regimes', Democratization 20, 13–38.

Goldstone, J. A. (2010). 'From Structure to Agency to Process: The Evolution of Charles Tilly's Theories of Social Action as Reflected in His Analyses of Contentious Politics', American Sociologist 41, 358–367.

Golob, T. and Makarovič, M. (2019). 'Reflexivity and Structural Positions: The Effects of Generation, Gender and Education', Social Sciences 2019 (8) 248; doi:10.3390/socsci8090248.

Goorha, P. (2010). Modernization Theory. Published in the Oxford Research Encyclopedia, International Relations, DOI: 10.1093/acrefore/9780190846626.013.266. Available at SSRN:https://ssrn.com/abstract=3412147.

Gray, M. (2019). Theorising politics, patronage, and corruption in the Arab monarchies of the Gulf, in de Elvira, L. R., Schwarz, C. H., Weipert-Fenner, I. (eds.). Clientelism and Patronage in the Middle East and North Africa Networks of Dependency, pp. 47–68. London and New York: Routledge.

Haddad, B., and Schwedler, J. (2013). 'Editors' introduction to teaching about the Middle East since the Arab Uprisings.' In PS - Political Science and Politics, pp. 211–216.

Haggard, S. and Kaufman, R. (eds.) (1995). The Political Economy of Democratic Transitions. Princeton, NJ: Princeton University Press.

Halpern, M. (1963). The Politics of Social Change in the Middle East and North Africa. Princeton, NJ: Princeton University Press.

Haseeb, K. El-Din (2013). 'The Arab Spring Revisited', in K. El-Din Haseeb (ed). The Arab Spring: Critical Analyses (London and New York: Routledge) 4–16. DOI: 10.1080/17550912.2012.673384.

Hay, C. (2002). Political Analysis. A critical introduction. London and New York: Palgrave Macmillan.

Härdig, A. C. (2015). 'Beyond the Arab Revolts: Conceptualizing Civil Society in the Middle East and North Africa', Democratization 22, 1131–1153.

Harik, I. (1990). 'The Origins of the Arab State System;' in Giacomo Luciani (ed.): The Arab State, Berkeley: University of California Press: 1–28.

Herb, M. (1999). 'All in the Family: Absolutism, Revolution, and Democracy in Middle Eastern Monarchies'. Albany, NY: SUNY Press.

Hinnebusch, R. (2006). 'Authoritarian Persistence, Democratization Theory and the Middle East: An Overview and Critique, Democratization', 13 (3), 373–395, DOI: 10.1080/13510340600579243.

Hinnebusch, R. (2014). Historical Sociology and the Arab Uprising. Mediterranean Politics 19, 137–140.

Hinnebusch, R. (2015). Globalization, democratization, and the Arab uprising: the international factor in MENA's failed democratization. Democratization 22, 335–357.

Hinnebusch, R. (2018). 'Understanding Regime Divergence in the Post-Uprising Arab States', Journal of Historical Sociology 2018 (31): 39– 52. https://doi.org/10.1111/johs.12190.

Heydemann, S. (2002). 'Defending the discipline: Middle East studies after 9/11', Journal of Democracy 13 (3): 102–108.

Heydemann, S. (2007). Upgrading Authoritarianism in the Arab World, Analysis Paper. Washington: Brookings Institution/Saban Center.

Hudson, M. C. (1977). Arab Politics: The Search for Legitimacy. New Haven: Yale University Press.

Hudson, M. C. (1988). 'Democratization and the Problem of Legitimacy in Middle East Politics—Presidential Address 1987', MESA Bulletin, 22 (2), pp. 157–171.

Humphreys, M. (2005). Between memory and desire. The Middle East in a troubled age. California: University of California Press.

Huntington, S. (1968). Political Order in Changing Societies. London: Yale University Press.

Huntington, S. (1991). The Third Wave: Democratization in the Late Twentieth Century. Norman, OK: University of Oklahoma Press.

Huntington, S. (1996). The Clash of Civilizations and the Remaking of World Order. New York: Simon & Schuster.

Jamal, A. (2012). Of Empires and Citizens – Pro-American Democracy or No Democracy at All? Princeton: Princeton University Press.

Jamal, A., and Tessler, M. (2008). 'Attitudes in the Arab World.' Journal of Democracy 19 (1): 97–110.

Jawad, N. (2013). 'Democracy in modern Islamic thought', British Journal of Middle Eastern Studies 40, 324–339.

Jessop, B. (2008). State Power: A Strategic-Relational Approach. Cambridge: Polity Press.

Kamrava, K. (2018). 'Hierarchy and Instability in the Middle East Regional Order', International Studies Journal 14 (4), 1–35.

Karl, T. (1990) 'Dilemmas of Democratization in Latin America', Comparative Politics 23 (October 1990): 1–21.

Kedourie, E. (1992) Democracy and Arab Political Culture. Washington, DC: Wash. Inst. Near East Policy.

Köllner, P. and Kailitz, S. (2013). 'Comparing Autocracies: Theoretical Issues and Empirical Analyses', Democratization 20, 1–12.

Kramer, M. (2001). Ivory Towers on Sand: The Failure of Middle Eastern Studies in America. Washington, DC: Wash. Inst. Near East Policy.

Lewitsky, S. and Way, L. A. (2002). 'Elections without Democracy. The Rise of Competitive Authoritarianism', Journal of Democracy 13 (2): 51–65.

Linz, J. and Stepan, A. (1996). Problems of Democratic Transition and Consolidation: Southern Europe, South America and Post-Communist Europe. Baltimore, MD: Johns Hopkins University Press.

Lipset, S. M. (1959). 'Some Social Requisites of Democracy: Economic Development and Political Legitimacy', American Political Science Review, 53 (1), 69–105.

Lucas, R. E. (2004). 'Monarchical Authoritarianism: Survival and Political Liberalization in a Middle Eastern Regime Type', International Journal of Middle East Studies, 36 (February 2004), 103–19.

Luciani, G. (1990). The Arab State. London: Routledge.

Luciani, G. (2009). Oil and Political Economy in the International Relations and International Political Economy, International Relations of the Middle East 81–103.

Luhrmann, A. and Lindberg, S. I. (2019). 'A Third Wave of Autocratization is Here: What is New about It?', Democratization, 26 (7), 1095–1113, DOI: 10.1080/13510347.2019.1582029.

Lust, E. (2009). Competitive Clientelism in the Middle East, Journal of Democracy, 20 (3), 122–135.

Lust, E. (2011). 'Missing the Third Wave: Islam, Institutions, and Democracy in the Middle East', Studies in Comparative International Development 46, 163–190.

Lust-Okar, E. (2003). Why the Failure of Democratization? Explaining' Middle East Exceptionalism. New Haven: Yale University.

Mahdavi, M. (2008). 'Rethinking Structure and Agency in Democratization: Iranian Lessons, International Journal of Criminology and Sociological Theory', 1 (2) December 2008, 142–160.

Makdisi, S. (2011). Autocracies, democratization and development in the Arab region. Dokki, Giza: The Economic Research Forum.

Makdisi, S. (2017). Reflections on the Arab Uprisings. International Development Policy | Revue internationale de politique de développement [Online], 7 | 2017, Online since 13 February 2017, connection on 30 September 2019. URL: http://journals.openedition.org/poldev/2280; DOI: 10.4000/poldev.2280.

Malmvig, H., Quero, J. and Soler i Lecha, E. (2016). The contemporary regional order. In Methodology and Concept Papers: Re-conceptualizing orders in the MENA region: The Analytical Framework of the MENARA project, edited by Eduard Soler i Lecha (coordinator), Silvia Colombo, Lorenzo Kamel and Jordi Quero.

March, A. F. (2015). 'Political Islam: Theory', Annual Review of Political Science 18, 103–123.

Minkenberg, M. (2007). 'Democracy and Religion: Theoretical and Empirical Observations on the Relationship between Christianity, Islam and Liberal Democracy', Journal of Ethnic and Migration Studies 33, 887–909.

Mitchell, T. (2003). The Middle East in the past and future of social science. UCIAS Ed. Vol. 3. The Politics of Knowledge: Area Studies and Disciplines. http://repositories. cdlib.org/uciaspubs/editedvolumes/3/3/

Moore, B. (1966). Social Origins of Dictatorship and Democracy: Lord and Peasant in the Making of the Modern World. Boston, MA: Beacon Press.

Norris, P. and Inglehart, R. (2004). Sacred and Secular. Religion and Politics Worldwide. Cambridge: Cambridge University Press.

Nugent, E. (2019). The Psychology of Repression and Polarization (December 13, 2019). World Politics, Forthcoming. Available at SSRN: https://ssrn.com/abstract=3090050 or http://dx.doi.org/10.2139/ssrn.3090050.

O'Donnell, G. A. (1973). Modernization and Bureaucratic-Authoritarianism: Studies in South American Politics. Berkeley: Institute of International Studies, University of California.

O'Donnell, G. A. (1986). Transitions from Authoritarian Rule: Comparative Perspectives. Baltimore: Johns Hopkins University press.

Platteau, J. P. (2011). 'Political Instrumentalization of Islam and the Risk of Obscurantist Deadlock', World Development 39, 243–260.

Przeworski, A. (2006). 'Democracy and Economic Development'. In Masfield, E. and Sisson, R. (eds.), Political Science and the Public Interest. Columbus: Ohio State University Press.

Przeworski, A. and Limongi, F. (1997). 'Modernization: Theories and Facts', World Politics 49 (January 1997): 155–183.

Ragin, C. (2008). Redesigning Social Inquiry: Fuzzy Sets and Beyond. Chicago: The University of Chicago Press.

Rakner, L., Rocha Menocal, A. and Fritz, V. (2007). Democratisation's Third Wave and the Challenges of Democratic Deepening: Assessing International Democracy Assistance and Lessons Learned. London: Overseas Development Institute.

Ramsay, K. W. (2011). 'Revisiting the Resource Curse', International Organization, 65, 507–529.

Rivetti, P. (2015). 'Continuity and Change before and after the Uprisings in Tunisia, Egypt and Morocco: Regime Reconfiguration and Policymaking in North Africa', British Journal of Middle Eastern Studies 42, 1–11.

Ross, M. L. (2001). 'Does Oil Hinder Democracy?', World Politics, 53 (3): 325–61

Ross, M. L. (2015). 'What Have We Learned about the Resource Curse?', Annual Review of Political Science 18, 239–259.

Rowley, C. K., and Smith, N. (2009). 'Islam's Democracy Paradox', Public Choice, 139, 273–299.

Rueschemeyer, D., Stephens, E. H. and Stephens, J. D. (1992). Capitalist Development and Democracy. Chicago: Chicago University Press.

Salame, G. (1994). 'Introduction: Where are the Democrats?' In Democracy without Democrats? The Renewal of Politics in the Muslim World, edited by Ghassan Salame, 1–22. New York: I.B. Tauris.

Saouli, A. (2015). 'Back to the Future: The Arab Uprisings and State (re) formation in the Arab World, Democratization, 22 (2), 315–334, DOI:10.108 0/13510347.2015.1010813

Sawani, Y. M. (2012). The 'end of pan-Arabism' revisited: Reflections on the Arab Spring. Contemporary Arab Affairs 5, 382–397.

Schlumberger, O. (2007). 'Arab Authoritarianism: Debating the Dynamics and Durability of Nondemocratic Regimes.' In Debating Arab Authoritarianism: Dynamics and Durability in Nondemocratic Regimes, edited by Oliver Schlumberger, 1–20. Stanford, CA: Stanford University Press.

Schmitter, P. and Terry Lynn Karl, T. L. (1991). 'What democracy is . . . and is not', Journal of Democracy 2 (3): 3–16.

Schneider, C. Q. (2009). The Consolidation of Democracy: Comparing Europe and Latin America. Abingdon. New York: Routledge.

Schwedler, J. (2006). Faith in Moderation: Islamist Parties in Jordan and Yemen. Cambridge: Cambridge University Press.

Schwedler, J. and Chomiak, L. (2006). And the Winner is . . . Authoritarian Elections in the Arab World, Middle East Report, 238, pp. 12–19.

Stepan, A. and Linz, J. J. (2013). 'Democratization theory and the 'Arab Spring', Journal of Democracy. 24 (2), 15–30.

Stepan, A. and Robertson, G. B. (2013). 'An 'Arab' more than 'Muslim' Gap', Journal of Democracy 14, 30–44.

Taylor, W. (2014). Military Responses to the Arab Uprisings and the Future of Civil-Military Relations in the Middle East. Analysis from Egypt, Tunisia, Libya, and Syria. London and New York: Palgrave Macmillan US.

Tessler, M. (2002). 'Islam and Democracy in the Middle East: The Impact of Religious Orientations on Attitudes toward Democracy in Four Arab Countries', Comparative Politics 34, 337–354.

Tessler, M., Jamal, A. and Robbins, M. (2012). 'New Findings on Arabs and Democracy', Journal of Democracy 23 (4): 89–103.

Valbjørn, M. (2015). 'Reflections on Self-Reflections – On Framing the Analytical Implications of the Arab Uprisings for the Study of Arab Politics', Democratization 22, 218–238.

Valbjorn, M. and Bank, A. (2010). 'Examining the 'Post' in Post-Democratization: The Future of Middle Eastern Political Rule through Lenses of the Past', Middle East Critique 19, 183–200. doi:10.1080/19436149 .2010.514469.

Valbjørn, M. and Volpi, F. (2014). 'Revisiting Theories of Arab Politics in the Aftermath of the Arab Uprisings', Mediterranean Politics, 19 (1): 134–136, DOI: 10.1080/13629395.2013.856185.

Wagemann, C. and Schneider, C. Q. 2010a. 'Qualitative Comparative Analysis (QCA) Fuzzy-Sets: Agenda for a Research Approach and a Data Analysis Technique', Comparative Sociology 9 (2010) 376–396.

Whitehead, L. (2002). Democratization: Theory and Experience. New York: Oxford University Press, 2002.

Wiktorowicz, Q. (2000). 'Civil Society as Social Control', Comparative Politics 33 (October 2000), 43–61.

Yom, S. (2015). From resilience to revolution. How Foreign Interventions Destabilize the Middle East. Columbia: Columbia University Press.

Yom, S. and Al-Momani, M. H. (2008). 'The International Dimensions of Authoritarian Regime Stability: Jordan in the Post-Cold War Era', Arab Studies Quarterly, 30 (1), 39–60.

Yom, S. L. and Gause III, G. F. (2012). 'Resilient Royals: How Arab Monarchies Hang On', Journal of Democracy 23, 74–88.

Frane Adam, Matevž Tomšič

# Roots and Manifestations of Populism in Contemporary Democracies

**Abstract:** This chapter deals with the recent outbursts of populism in contemporary democracies. It determines the relationship between populism and the personalization of politics and discerns commonalities and differences between different types of populism. The thesis is that populism is a reaction to the crisis of democracy as well as to the anomalies and unintended consequences of globalization. We show that populist leaders are accentuating the loss of national sovereignty, the dominance of supranational institutions (including the EU), as well as the penetration of big capital and multinational corporations. However, despite some common ideological and cultural values, there is variety in its manifestations throughout Europe and beyond.

**Keywords:** populism, globalization, neoliberalism, nationalism, democracy, European Union, person-based politics

## Introduction

Populism in politics is currently a burning topic. However, it is not just about style and rhetoric, emphasizing a community based on national identity and opposing supranational institutions, nor is it some kind of a deviation from 'true' democracy. Whichever opinion we hold about this phenomenon – represented by leaders such as Trump in the USA, the events related to Great Britain's withdrawal from the EU (Brexit), Orban in Hungary and Kaczynski in Poland, or outside of the EU like Putin in Russia and Erdogan in Turkey or Chavez in Latin America[1] – we to recognize and admit that now, more than ever, the limits of globalization in the sense of free trade (i.e., the free movement of people, goods and capital), as well as transnational integration are under serious challenge (cf. Russell Mead 2017). Ironically, in 2016, mass protests took place in (Western) Europe – mainly under the aegis of the Left – against the trade agreement

---

1 Some authors distinguish between right-wing and left-wing populism. The latter is expressed in Southern Europe (Podemos or Syriza in Greece) and mainly in Latin America which has also a rich history of populist politics like Peronism in Argentina Rodrik 2017; cf. Laclau 2005). However, there is not agreement which parties or movements can be considered as populist, Mueller (2016 b) rejects the statement on populist character of Syriza and Podemos while Mouffe (2016) argues the opposite.

between the USA and the EU (TTIP). After that, came the rapid rise to power of the right-wing Donald Trump, who would abolish all of these agreements without even intending to start new ones. All of this is in the name of the alleged damage that such agreements would cause American workers and American national interests. The rise of populism in contemporary democracies is the entanglement of both structural conditions and the characteristics of political actors. It is related to the change in the nature of social cleavages, prevailing value patterns, modes of political organization and decision-making as well as to the (lack of) responsiveness of established political elites and their (in)ability to resolve critical social problems. It is a major social process, inviting theoretically informed analysis of societal transformations (Rončević and Makarovič 2011; Rončević and Makarovič 2010).

The thesis we are putting forward is that populism has skilfully taken advantage of the inability of the mainstream Left and Right to articulate and implement a programme for regulating world trade, migration, the financial industry, tax havens and the grossly overextended role of multinational corporations. In short, contemporary populism is a reaction to the anomalies and (unintended) consequences of globalization (Rodrik 2017). However, whether this reaction is adequate or whether it brings any solutions to these anomalies is highly questionable.

The aim of this contribution is not the historical and theoretical elaboration of the meaning of the populist phenomenon in politics. However, some preliminary description of its characteristics is needed. Populism, as Rodrik (2017) says, is only a loose label pointing to a diverse set of movements and regimes. They have a common denominator that contains anti-elite and anti-establishment stances (Mudde and Kaltwasser 2017), a claim to speak for the people against elites, opposition to neoliberal economy and globalization as well as, in some cases, a preponderance toward authoritarian rule (Rodrik 2017, p. 1–2). According to another recent author, all types of populisms show a certain hostility toward pluralist society (Mueller, 2016). In contrast, populism is an extreme consequence of the personification of politics (Tomšič and Prijon 2013) and the arrival of strong leaders (from Berlusconi in Italy some years ago to Orban today in Hungary).

However, as already noted, our focus is on populism as a reaction to the issues and anomalies of globalization such as free trade, migration and the enlarged influence of massive techno-economic corporations. While, according to some authors, free trade is destroying workplaces in specific sectors of the economy both in the EU and the USA (Damijan, 2017),[2] global migration flows have been

---

2   One recent study analysed how global trade and technological advancement have influenced the labour market in the USA and EU. Their conclusion is that, between

met with inadequate integration measures and consequent cultural conflicts. Large (multinational) corporations and financial institutions continue to actively pursue various channels for tax evasion (tax havens), handsomely reward their managers and owners, thereby contributing to the rise of social inequality

Yet there are differences between populist leaders. Orban and Kaczynski might have a close relationship but, regarding their attitude to Putin, they are not in the same boat. Further, Poland and Hungary are highly unlikely to exit the EU (which was one of the key parts of Le Pen's platform in the last election; she, too, like most populist leaders, presumably including Trump, has close relations with the Kremlin), because both depend on the financial resources that they draw from EU structural funds.

One could say that globalization, in conjunction with technological advancement over the previous two decades, has altered the (social) character of both developed and underdeveloped societies. Populism can hence be viewed as a signal of political transformations as well as a signal pointing to the more profound transformations of social relations and modes of integration. The big question of how to regulate globalization processes to reduce the negative and increase the positive effects thus remains unanswered. Individual states appear to be helpless, although populist leaders tend to spread the illusion that closed national systems can provide better shelter against global issues. Nevertheless, the EU could surely benefit from a more active approach and by utilizing its full knowledge potential regarding its role and significance in defining the course of international relations and the future developmental model. However, we know that in its present form, the EU is incapable of focusing on such delicate and

---

2001 and 2011, free trade institutionalised in the form of the WTO – together with technological restructuring which also a part of the globalization process – contributed to the loss of 5.7 million jobs in industry in the USA and even more in the EU countries – 7.2 million. According to the authors, mostly affected were middle qualified workers, while the less qualified and managers benefited (Beemersch et al. 2017). In contrast, one of the neoliberal opponents of Trump's current politics of protectionism states that free trade was not at all responsible for the employment loss in American industry. However, this is only a statement without any statistics and empirical evidence. The author is an American academic economist; it seems he primarily speaks in this case rather as an ideologue than as a scientist (see Irwin 2017); other analyses show that global free trade and import penetration from China (after 2001) deeply influenced the situation on labour market and even contributed to the election victory of Donald Trump (Author et al. 2016).

complex issues. Therefore, it is not just reforms that are needed, but an elaboration of new concepts and intellectual revitalization.

Regarding the structure of this text, first, the often opposing views of populism will be presented, following which the three models of societal development with their emphasis on the autonomous (auto-centric) and dependent types of socio-economic (international) interrelations will be explained. In terms of Donald Trump's electoral success, as well as Victor Orban's in Hungary, the essence and future of populist anti-globalist ideology will be illustrated. Finally, the European Union's role as a more pro-active global player in regulating global processes is underlined.

## Populism and erosion of the left

Many see populist parties and movements as connected with (proto-)fascism. Others seek to use more analytical ways to describe this phenomenon from different points of view. It seems that a better approach would be to understand the specific features of contemporary populism. Its ideology is not always consistent, and it acts on various levels. The dilemma of how to interpret or formulate conclusions in this regard is also present in a special issue of the leading Leftist-oriented Slovenian magazine Mladina dedicated to populism. Some authors insist that populism is the first step to a fascist regime, while other authors are more cautious in defining, or respectively resisting, any immediate stigmatization of populism (Zupančič Žerdin 2016–2017; this author is a close collaborator of Slavoj Žižek; Rizman 2016–2017).

In contrast, the French philosopher Alain Badiou maintains that we are dealing with 'democratic fascism'. Populist leaders may indeed act within the democratic constitution, but they act outside it, and their actions lie beyond it (like Mussolini and Hitler in the 1930s), thus justifying the label 'fascist' (Badiou 2016–2017).[3]

A completely different, wholly affirmative attitude to this phenomenon is expressed by the Slovenian sociologist Tomaž Mastnak. His thesis is that neoliberalism, not populism, is closer to fascism and the only real opponent of neoliberalism, cosmopolitanism and anti-nationalism is populist politics. This is how

---

3   *Inter alia*, Badiou collaborated with another well-known philosopher, Slavoj Žižek. Today, they hold very different, even opposing, opinions on populism. It is known that Žižek has expressed some understanding for Trump, namely that he would shake-up the establishment and disrupt the *status quo* (Žižek 2016). His opinion is also mentioned in Zupančič Žerdin 2016–2017.

he concludes his discussion: '[o]pposing populism means opposing democracy. The political choice, which we face today, is a choice between liberalism and democracy. In the given situation the choice is: either neoliberalism or populism' (Mastnak 2016–2017, p. 47).

The Left, Mastnak claims, is mainly on the side of the neoliberal elites. He states that multiculturalism, inclusivism, cosmopolitanism, genderism, LGBTism, xenophilia and Islamophilia 'are the death of the Leftist political language'. The Left – both the social-democratic and the more radical – is dysfunctional and may be doomed to disappear. This is the harshest appraisal of the Left's role – and it comes from under the pen of an author who used to define himself as a radical Leftist. The only alternative to neoliberal global domination and the dismantling of democracy is a return to economic nationalism, national sovereignty and the rule of the people (not the elite) – and this is the programme of populist parties. It seems that in international (radical) Left circles, the majority rejects the notion of populism. However, there is an influential group with a different approach that proclaims left-wing populism to be the most suitable strategy to reach the ideological hegemony, in Gramsci's sense of the phrase.[4]

---

4    In this connection, the (post)Marxist political philosopher Ernesto Laclau can be mentioned. Long before recent wave of populism, he wrote a text on populism in connection with capitalism and fascism (Laclau, 1977) and later a book (Laclau, 2005) with the clear intention of rehabilitating this term as politically relevant by saying: 'Populism is, quite simple, a way of constructing the political' (Lacla, 2005, p. xi). He is dedicated to the question of how to use the mobilising effect of populism to broaden the basis for socialist hegemony. Namely, the mobilization effect of class struggle is too weak to reach the sufficient power in this regard. This is one of the possible interpretations of his not very clear-cut argumentation in the book. The following quotation seems to be instructive: 'The return to the "people" as a political category can be seen as contribution to this expansion of horizons, because it helps us to present categories – such as class – for what they are: contingent and particular form of articulating demands, not an ultimate core from which the nature of the demands themselves can be explained. This widening of horizons is a precondition for thinking the forms of our political engagement in the era of what I have called globalised capitalism' (Laclau, 2005), p. 250; a more understandable and concrete explanation of the 'chain of equivalence' can be found in an interview with his close collaborator, Mouffe. She says: 'In response, the Left must create what I call a "populist frontier" of all the popular classes against the elites and establishment' (Mouffe 2016).

   One of the fiercest of his opponents of Laclau in that time (2005) was Žižek, who strongly rejected any idea that populism in itself is not linked necessarily to a certain ideology, labelling it as proto-fascism (Žižek, 2006). As seen recently, Žižek has latter

Mastnak's position is much more radically elaborated and freed from any Leftist sentiment. Interestingly, he (a long-standing visiting professor at the University of California, USA) has published several similar contributions in the main (left-leaning) Slovenian newspapers Delo and Dnevnik (where he has a regular column). In them, he expressed support for Trump's election as well as previously standing up for Victor Orban and the Kaczynski party in Poland. In one of his columns, he analysed the presidential elections in France and expressed his preference for Marine Le Pen. Surprisingly, these claims and analyses provoked almost no objections or controversy among neither his former like-minded adherents nor in the broader intellectual public. We believe that Mastnak's analyses help us understand the essence of populism and the dynamic restructuring of the political space. However, it is more difficult to accept his simplified relationship between liberalism, populism and democracy, in particular his statement about the nonselective closure within national frameworks, including withdrawal from EU membership.

In many European countries, the erosion of the Left as a consequence of the populist ideological offensive is actually taking place. This is particularly true for Hungary and Poland; however, in France and apparently in the Netherlands, Germany and Austria it seems to be losing ground.

## The closedness and openness of national systems

In terms of ideal types, the sociological literature distinguishes three models of development. The first is the auto-centric model based on closure, the second is the auto-centric model based on openness, while the third is the dependent-peripheral model (Menzel and Senghaas 1986). Autocentricity means that the social or national system controls the resources and the ability for self-regulation by making critical strategic decisions, which serves to maintain the boundaries and identity of the system. However, there is a substantial difference between the open and closed auto-centric models. The auto-centric model based on openness is a combination of endogenous developmental factors and actors, and of integration into the international environment, including, of course, a certain dependence on that environment.

Such societies need to be meritocratic because they have to activate all their knowledge potential and develop high-quality institutions in order to be able to come to terms with the demands of a more complex environment. The

---

changed his opinion (or theory). It is difficult to say that he is supporter of populism, but he showed some understanding for Trump during the campaign in 2016.

auto-centric model based on closure seeks to reduce its dependence on the environment while stressing national sovereignty and self-sufficiency. In the modern globalized world, such a model is, of course, unrealistic. We can at best talk about a tendency towards lesser or greater closedness/openness. If we connect this framework to the current debate, then we can say that populist politics tend towards greater closure with regard to the model of auto-centric development based on (relative) closure. Trump's programme is typical here: he would like to build a wall along the Mexican-USA border and abandon trade agreements or other international commitments (especially environmental ones and those connected to climate change).

The main feature of the third model – the peripheral-dependent – is that it is too open to the environment, thus outside interests dictate the major decisions, and domestic actors lose their capability for independent operations. In the more extreme cases, this is the so-called neo-colonial situation, whereas, in a milder version, it is a model in which domestic actors have no proper strategy to align foreign capital or foreign policy interests with national priorities. The populist ideology emerges precisely from the feeling that their countries have strayed onto the path of peripheral development, that they have lost the attributes of a strong state and are now unable to face external pressures. Brexit was born out of such feelings. Moreover, in the Eastern part of the EU, the opinion of being subordinated to the EU, especially the loss of auto-centricity, is spreading – inevitably leading to populism and state-led capitalism.

## Neo-colonialism, globalism and nationalism

Some social scientists speak of a neo-colonial relationship between the Eastern and Southern EU member states in relation to the core EU states. This has often been voiced in connection to Greece. An influential group within the framework of the so-called Social Anthropology of Transition is especially loud. However, since this group is mostly composed of Neo-Marxists, they are not particularly enthusiastic about nationalism and populism. They are waiting for the anti-capitalist revolution or for the emergence of the 'Great Man', as the Slovenian anthropologist Vesna Godina puts it in her book (2014). It seems that the 'Great Man' is already closer to a populist leader.

Victor Orban garnered such great support in Hungary because he was able to articulate people's feelings about the fact that, as a result of the reign of their Left or Social-democratic party (the former Communists), the country has lost its attributes of sovereignty, has been flooded by foreign capital, and hence become dependent on large foreign corporations and big capital. This is also supposed to be the fault of the EU, namely Brussels; the conclusion is that it is necessary to assert national interests in this regard, even at the cost of a dispute. With the migration crisis, all of this has further deepened and somehow managed to unite the Visegrad countries. However, the situation is similar in the West, where one of the main reasons for Brexit was to exercise control over migration flows and, for a large proportion of the public, immigration from the Eastern EU countries was just as controversial.

The well-known Czech economist and philosopher Tomáš Sedláček[5] sees an irreconcilable contradiction between national identity and the openness of globalism (see his interview in Delo, 10 May, 2017, p. 3). For him, one is either a closed nationalist or a globalist committed to openness. There is no middle position. Sedláček mentions that the Czech Republic has not fared the best in the migration crisis, but he nevertheless believes his country will soon join the leading group of countries in the EU. However, despite being an analyst for a large commercial bank in Prague, he fails to explain (as on previous occasions) why the Czech Republic refuses to adopt the euro. If he had conducted this particular analysis, he would probably no longer think in terms of either-or categories. Namely, this issue is not just relevant to his country since Sweden and Denmark also refuse to join the eurozone. This is certainly an element of closure, that is, of exercising national sovereignty at the expense of Europeanization.

However, we cannot say that either the Czech Republic or the aforementioned Scandinavian countries are economically or otherwise self-sufficient, but they do want to keep their hands on the levers of their monetary policy. Moreover, they are not doing badly either. Two years ago, the Czech Republic's GDP, in terms of purchasing power, overtook Slovenia's, while the other two countries, as we know, are some of the most advanced in the world. This should not be seen as a recommendation to follow them but to contest Sedláček's thesis about the exclusionary relationship between systemic closedness and openness. In actuality, there are trade-offs and different combinations of closure and openness here.

## Personalization of politics

The rise of populism is related to the profound change in the constitution of political space that has been taking place in contemporary democracies in recent

---

5    Sedláček is the author of a book published in English (and translated into other languages as well) that for a while was a bestseller (Sedláček, 2012). The gist of it is that economics should not be so technical and should become more open to the humanities and philosophy.

decades, which was particularly applicable to the developments in terms of the organizational structure of political parties and their mode of functioning as well as the change in mechanisms of mobilization of political support and establishing a link between parties and their constituencies. We are talking – above all – about the 'personalization of politics'. Although some authors claim that empirical evidence of this phenomenon is mixed, at best, (Kriesi 2011), it is hard to deny that personal traits of political leaders play an increasingly important role in the political life of contemporary democracies (Van Zoonen and Holz-Bacha 2001), which is called by some the 'presidentalization' of politics' (Poguntke and Webb 2005; Passarelli 2015; Webb et al. 2012).

The importance of social cleavages, meaning class, religious and regional divisions that crystallized during the process of modernization of Western societies that constituted the basis of the formation of modern political party system (see Lipset and Rokkan 1967), diminished over time and had 'no longer the same hold over European population as it had been the case previously' (Blondel and Thiebault 2010, p. 1). Among the main factors that contributed to this, one can state the value change observed by some researchers (see Inglehart 1990) to take place in developed societies decades after World War II, which resulted in a decrease of attachment of the people to classical identities such as class, religion, ideology and similar. This was combined with a decrease of confidence in traditional political institutions in the vast majority of established democracies (Diamond and Gunther 2001). Consequently, the position of political parties that used to be described as organizational vehicles of collective action and social choice (Kitschelt 2001) and channelling agencies that represent a link between society and government institutions (Sartori 1976) has weakened.

The power of a political leader is related to his/her control over the political process in the party, which includes the party programme and policies, the appointment of party executive personnel and selection of party candidates for electoral positions, as well as the strategic decisions of the party (e.g., with whom to form a coalition, whether to join the government or remain in opposition, etc.) (Blondel and Thiebault 2010, p. 69). In this way, the leader exerts prevailing influence over all key elements of the party's conduct. However, personalized leadership does not appear in 'empty space', since it is the party system itself, of a particular country that represents the framework within which it 'emerges with significant support in the population in the context of a liberal democratic system' (ibid. 261).

The strengthening role of the party leader also applies to the development of political parties, as such. There are evident changes in terms of the structure and mode of functioning. One of the most important ones is the decline of party membership that has been taking place in recent decades. As stated by van Biezen and Poguntke (2014, p. 205), 'parties are struggling to hold on to their membership organizations and are failing to recruit significant numbers of new members'. This is related to the above-mentioned diminishing importance of collective identities and corresponding weakening political loyalties. Due to this, citizens less and less see political parties as the framework of their social engagement.

In contrast, the functioning of political parties decreasingly relies on its members. Many activities are now 'outsourced': delegated to professional organizations and teams with specialized expertise (public relations agencies, think-tanks, pollsters, etc.). Contemporary politics became 'more and more about the competition between professionalized party elites and less about the mobilization and integration of socially distinct groups' (ibid. 206). As a result, links between party elites and 'ordinary' members progressively weakened. There are even parties that do not need party members to take part in the political process (as is the case with Wilders' Party for Freedom in the Netherlands) (see Mazzoleni and Voerman 2016). In this regard, one has to mention the new types of political parties like 'cartel party' (Katz and Mair 1995) and 'business firm party' (Hopkin and Paolucci 1999; Krouwel 2006).

The personalization of politics is related to the rise of so-called 'niche parties' (sometimes labelled 'single-issue parties'). Parties of this type reject traditional the class-based orientation of politics, transcend socio-economic cleavage and are – unlike traditional 'catch-all parties9 – focus on a narrow set of predominantly non-economic issues (Meguid 2007; Wagner 2012). They often follow extremist or at least non-centrist ideologies. To this group belong parties of ecologist, regionalist or ethno-nationalist orientation. Although classical political parties can become 'personalized', the role of personalized leadership is much stronger in the conduct of 'niche parties'. In recent years, we have witnessed the appearance of a new type of 'niche parties' without a coherent ideology and political programme, which are not focused on particular policy-issues but are generally oriented against the existing political establishment, that is, against the 'old' political elites. The most evident and the most politically successful example in established democracies is the Five Stars Movement in Italy. 'Niche parties' are even more present in the new democracies from Central and Eastern Europe. They more often than their Western counterparts lack clear programme and policy orientations. Instead, they build their public appearance on general criticism of established political parties and the 'character' of their leaders. The irony is that such party can easily transform themselves into a mainstream one; examples include Direction in Slovakia, GERB in Bulgaria, ANO 2011 in the

Czech Republic, or Party of Modern Centre and List of Marjan Šarec in Slovenia. However, this phenomenon is not limited to the new EU member countries, as proved by the stunning success of Emanuel Macron and his 'neither-left-nor right' party La Republique En Marche at French presidential and parliamentary elections in 2017.

There were personalities of different political leaders that in recent decades have decisively contributed to the emergence of new political 'competitors', which refers particularly to parties of regionalist or autonomist orientation and also to those on the extreme right – politicians like Salvini, Le Pen, Strache, Wilders, among others – but also to some parties that become a part of the political mainstream in their countries – as it is the case of Silvio Berlusconi and his party. However, some leaders actively contributed to the renewal of traditional parties. In this regards, the most evident example is the role of Tony Blair in the rise of New Labour.[6]

Personalized politics brings a different style of political communication to the forefront. Its main accent is on the presentation of personal characteristics of the particular political leader – his strength, energy, vision, organizational skills and similar. The marketing strategy is focused on his charisma much more than on his political vision, values and party programme. The electoral campaign is based on establishing a leader's close contacts both with the electorate (ordinary people) and media. The whole promotion is concentrated on events in which the leader is playing the central (or even the exclusive) role. Other persons in the party play a secondary role, serving mostly for his support. The leader is the point through which all essential political messages are transmitted to the public and especially to his party's electorate.

This personalized type of political approach rests both on strong 'face-to-face' contacts of the leader with his supporters, as well as on the intensive use of media communication. These politicians are very keen to present themselves as 'men of the people' who have a deep understanding of the needs and wishes of ordinary citizens.

There are some personal traits common to most of the 'non-partisan' politicians. They are, as a rule, strongly extroverted personalities who are prone to practice direct contact with the voters. Simultaneously, they are very skilful

---

6    Recently, the same party again experienced strong personal leadership under Jeremy Corbyn who positioned it more to the traditional left. However, this attempt was not successful since the Labour Party experienced painful defeat in the 2019 parliamentary elections.

in media communication, meaning that they are able to establish an appealing image in public through the use of visual impressions. They have strong charisma by making the followers believe that they can 'make things better'. Their discourse addresses people of a certain country or a region of a city as a whole rather than a particular social group or a constituency based on a particular ideological platform. This approach is often characterized by strong populism appealing to ordinary people and claiming to share their thoughts and sentiments. Non-party politicians tend to present themselves as political 'outsiders' who do not have anything in common with the established political structures and who express the genuine will of 'ordinary people' who are supposed to be mistreated by the old elites.

## Lack of leadership and dissatisfaction with established elites

This development came about in a situation characterized by the poor performance of the established political parties and their governments. The low administrative efficiency of governments in many European countries was accompanied by a lack of responsibility (Tomšič and Prijon 2013; Tomšič 2017a), which actively contributed to the weakening of confidence in politics and politicians. The poor image of political institutions, especially political parties, regardless of their ideological orientation, became predominant in the assessments of the population (see, for example, Newell 2010). The trend of decreasing trust in political institution is evident in many Western democracies but is more profound in the new democracies. Among them, political parties are among the most distrusted institutions (Makarovič and Tomšič 2015). There are many elements related to the behaviour of established political parties, such as ideologization, clientelism, corruption and other dysfunctional practices that contribute to such negative attitudes. An increasing representation gap between parties and electorates exists in many democracies (Keman 2017). In such a climate, 'new faces' can easily gain popularity, especially those who build their campaigns on personalized and sometimes 'non-political' platforms. They have been building their campaigns either through the 'managerization of politics', based on the notion of 'politics as business' according to which the country should be run as a company (the case of Andrej Babiš, the winner of the last Czech parliamentary election in 2017), or through the 'moralization of politics', which proclaims a moral renewal of politics that brings higher standards of political culture (the case of Miro Cerar, the winner of the Slovenian parliamentary election in 2014) (Cabada and Tomšič 2016).

There is a widespread perception of inefficiency of democratic political institutions and of a lack of leadership reflected in the incompetence and irresponsibility of established political elites. The protests (often violent) in different parts of the European Union that have been taking place in recent years (in some countries – France, Greece, Hungary – they appeared even before the global economic crisis) were manifestations of the discontent of the citizens (or at least some social groups) with the current socio-economic and political situation. Poor coping with first the financial and later the migrant crises at both national and European levels strengthened these feeling. In particular, the latter played a critical role. It turned out that the Union has no scenario for how to effectively deal with the great mass of people from its nearer and more distant surroundings who wish to settle within its borders. The poor performance of both European and national institutions in dealing with migration issues gave a strong impetus for populist political forces. They campaigned with the presentation of themselves are defenders of their home people against foreign 'intruders', on the one hand and against 'treacherous' established elites on the other.

However, not only the performance of established elites is a reason for increasing dissatisfaction among the citizenry. Their (alleged) value orientations are also a target of criticism and object of increasing rejection among them. Certain ideologies are advocated by at least a part of the elite. This particularly refers to multiculturalism, which can be seen both as policy-strategy (how to settle relations between different culturally specific entities, i.e., ethnic, religious communities) and as an ideology (promotion of positive nature of intercultural differences). Here we are focusing on the second aspect since it is the one that is particularly related to the rise of populism. As an ideology, multiculturalism rests on the notion that cultural diversity is something that is almost inherently positive. It claims that individual culturally specific communities must have the right and the opportunity to cultivate their values, customs and lifestyles (Heywood 2012). It argues for equality among these communities, focusing on the rights of minority communities vis a vis cultural majority (for example, immigrant communities within European societies).

Multiculturalism is related to the rise of post-materialist values (Inglehart). It is mostly endorsed by members of the academic community and other opinion-makers as well as by part of the political elite. Some even perceive it as something of universal value. However, the problem of this ideology is in its downplaying of the relevance of (too wide) cultural differences and their potentially problematic impact on the functioning of society (Tomšič 2017b). With the migrant crisis and the problems it brought, these ideas were met with wide opposition throughout Europe. Many people blamed the multiculturalism of (a part of) elite as the reason for the poor handling of migration-related problems.

Moreover, the right-wing populists were the ones that took advantage of these sentiments, accusing established elites of being responsible for these problems. They have been claiming that, based on multiculturalist ideology, the elites became alienated from the needs and wishes of ordinary 'autochthonous' people. This led many of these populists to align with strongmen like Vladimir Putin, who is perceived as a traditionalist and, as such, a stringent opponent of multiculturalism.

In relation to multiculturalism, one should also mention other elements of ideological discourse of the contemporary Left like advocating interests of the LGBT community, opposition to traditional family values and the promotion of the idea of gender as a purely social construct (meaning there are not only two but many different genders). Again, these elements are rather well endorsed within (a certain part) of elites but firmly rejected by a majority of citizens. Populists instrumentalize this situation, picturing it as another proof of the 'decadence' (or even 'degeneration) of the established elites and their value systems, and proclaiming themselves as tough defenders of 'genuine' (i.e., traditional) values and moral orientations of the people.

## Populism and the new media

The rise of a political approach called 'Berlusconization'[7] (Mancini 2011) by some is related to several international developments in contemporary societies. One of them is an increased role for the media in the political process, called the 'mediatization of politics' in which modern mass media, especially electronic ones, increasingly build their stories on 'spectacle' in which images play a more important role than ideas and programmes do (Campus 2010). In such circumstances, how a political candidate 'presents' him/herself to the audience (supporters, voters) is more important than how relevant and/or feasible are his/her political proposals and solution to resolve political, economic and social problems.

One should not neglect the role of new digital media in the rise of populism and the personalization of politics. These media are based on network principles,

---

7   The phenomenon of leadership style known as 'Berlusconisation' is based on the person of the Italian media magnate, politician and former Prime Minister Silvio Berlusconi, who represents a new model of politics that can be identified in some contemporary democracies.

erasing the border between producers and users of information, which leads to a dehierarchization of the news production process in which the role of traditional gatekeepers is reduced. The use of social networks allows political messages to be spread beyond traditional media channels (as the example of Donald Trump's presidential campaign shows; he managed to 'bypass' traditional media – which were mostly adverse to him – through Twitter and other new media). The use of the Internet enables direct communication between a leader and the people (the constituency). Although new media allow more and more people to spread their ideas and views, there has been an increasing amount of discussion in recent years about the negative aspects of their use. Many see social networks as a significant generator of spreading misinformation, 'fake news' and 'hate speech'. This ought to be linked to the increasing dispersion of information resources and to this lack of control. These negative media phenomena are often perceived as an integral part of populist discourse (Mudde 2004).

However, person-based politics is not solely a trait of populism since the politics of the present day has likewise gone through many changes, from mediatization and adjustments to media portrayals through to the increasingly greater role of public relations. Much of modern politics is currently in the limelight and the adulation of specific social groups as well as the promise and forecasts of what one strong leader (or a party led by a strong persona) can accomplish for the people.[8]

---

8   Interestingly, the *enfant terrible* of Austrian politics, the leading Austrian politician Sebastian Kurz run at the 2017 election on his own ballot (The List of Sebastian Kurz – Austrian People's Party). Although already a member of established conservative party, his name is at forefront, and the party represents only the formal framework. In this connection, Boris Johnson, current British Prime Minister and the 'brain behind Brexit', should be mentioned. French President Emanuel Macron is another representative of the personification of politics (although he is not a populist, he is actually praised for defeating Le Pen's populist *National Front*). It is known that the former leading party of Slovenian government is named after its leader and former Prime Minister Marjan Šarec (List of Marjan Šarec) who, as a new face in politics and a number of populist elements, was able to form government coalition after 2018 elections (although his party did not win relative majority) which lasted until March 2020. Also, the president (Borut Pahor) is known for skilfully using populist rhetoric and approaching to the 'little men' during campaign.

## How and why a neoliberal fortress became a populist heaven

Trump's electoral victory by way of mobilizing the so-called losers of globalization along with the British referendum decision for 'Brexit' are somewhat surprising since both (Anglo-Saxon countries) represent the most typical form of so-called Liberal Market Economies (LMEs) and neoliberal politics. There are politicians like Boris Johnson who are not typical populists but have the potential to transform into such. There is also the question of how long Trump's government will survive and be able to put to rest the serious allegations, related to the 'Ukrainian scandal' (which is the basis for ingoing impeachment procedure).[9] He also encounters opposition from various parts of society. However, it can be said that Trump is the most typical representative of populist politics in the contemporary world. In any event, the specific American context (as well as other national contexts in which populists play an essential role) should also be taken into consideration.

If we consider Trump's public appearances, we can state that they are a mixture of trade unionism, a missionary approach and robust simplification. One rally in Pennsylvania in late April 2017 was very telling and also interesting from the perspective of studying the iconography and distribution of Trump's followers (respectively, of the closest invitees) who formed his escort. At least two representatives of the working class were present. One of the major emphases of this congregation was the promise that the new government will reopen coal mines and provide work for 48,000 unemployed. What the environmental consequences will be of renewing the operation of the mines as well as the use of coal for power, of course, went unmentioned. Trump, furthermore, promised new jobs in companies that he would supposedly convince not to relocate their operations to other countries. How he would achieve this remains unclear. The plans for his infrastructure projects likewise have yet to be seen (cf. Paulson 2017).

Another highlight is the fact that the trade agreement between the US, Mexico and Canada cannot simply be discontinued, but Trump aims to re-negotiate it and achieve an adjustment of the agreement to favour the national interests of the USA. One of the most common words he used was 're-negotiate', which recently refers particularly to USA trade battles with China.

Speaking of the wall along the USA-Mexican border, what comes to mind is a book written by an American economist and a British political scientist

---

9    It revolves around pressures, exerted by U.S. President Donald Trump on Ukrainian
     authorities in order to provide him with damaging information about 2020 Democratic
     Party presidential primary candidate Joseph Biden.

entitled Why Nations Fail? (Acemoglu and Robinson 2013). The central thesis is that successful socio-economic development necessitates quality and inclusive institutions as opposed to extractive institutions that hinder development. The first chapter mentions the city of Nogales, Arizona, which is divided into American and Mexican parts. The authors show how the American part of town is an example of prosperity, while the Mexican part, despite the city being located in the most prosperous part of Mexico, still lags considerably behind the American urban area. The problem emerging today is that Trump wants to build a wall that would separate two democratic countries. Or does organised crime – which has reached truly grotesque proportions in certain parts of Mexico – disregard all democratic institutions? Another question is how these institutions function in anomic and clientelistic societies such as Mexico. Whatever the case may be, the authors of the mentioned book give a very schematic and also quite useless theory of development. Above all, they are uncritical of the situation in the United States. The victory of Donald Trump is likewise a result of shifts in the character of society since American society is losing its meritocratic orientation and is approaching a plutocratic society (a society in which a handful of super-rich rule), which means it is not self-evident that inclusive institutions still dominate in the USA, as the authors claim.[10]

## The future of populism

Over the last two decades, the world has experienced many changes in its social structure, values and international relations. The effects of globalization are contradictory: for some countries, regions or social groups they are beneficial while for others they remain negative. If Trump had not won in the United States, we would possibly not even know that elections in that country can be won by mobilizing both the losers and opponents of globalization. We had all previously

---

10 On the last page, they describe Chinese society as a society based on extractive institutions that hinder growth. However, the question remains of how to explain its high growth rate in the last few decades. They simply say that also countries with extractive institutions may experience short-term growth, like the Soviet Union in the 1970s. With regard to China, it can be said that this country is a unique case and that the authors do not respect that. It has a one-party system, the Communist Party is hence very powerful but, on the other side, we have to do with a mixture of state capitalism and powerful crony capitalists. It seems that China is economically better off due to its skilful adaption to globalisation and its certain respect for meritocracy (Bell, 2015).

been convinced that the United States was the primary driver of globalization, yet it turns out that this is not entirely so. We know – and this is also said by the aforementioned Sedláček – that China is now the state that is most interested in the smooth operation of free trade and the movement of capital (as also evident at the meeting in Davos). However, this is not the same as openness (as claimed by the same author). Although China does, to a certain extent, take meritocratic principles into account, it has yet to become a democratic society (the thesis about the meritocratic nature of this country is advocated by the US sinologist and political scientist Daniel A. Bell 2015). It is not open internally, and it is not an open society (Shirk 2017).

Considering only the situation in Europe, there are significant differences with regard to the strength of populism and its impact on political life. In some countries, populist parties are the sole power-holders as is the cases with Hungary (Fidesz) and Poland (Law and Justice). In other countries, populist parties assume the role of leading partners in government as is the case with Babiš's ANO 2011 in Czechia or Šarec's party in Slovenia, or minor partners in government what was the case with Freedom Party in Austria. Italy represents a specific example since in 2018–19, it had a coalition of two populist parties/ movements in power (League and Five Star Movement). Even when populists are not in power, their political weight varies significantly, from representing strong opposition (National Front in France or Party of Freedom in the Netherlands) to being weak or almost non-existent opposition (Ireland, Portugal).

The European Union should be more interested in the greater regulation of the disordered and anarchic globalization processes and assert itself more as a global actor in terms of an auto-centric model based on openness. This does not nec- essarily mean a trade war and brute protectionism, as Trump imagines. It does, however, mean that global trade needs to be in the function of sustainable devel- opment and marked by the quality of consumer goods, high safety standards and consumer rights. This would limit the currently excessive low-quality consum- erism and the wild proliferation of shopping malls (in this regard, Slovenia is neoliberal par excellence and even at the forefront of the EU).

As far as populism is concerned, while it may never become established as a dominant political option, it will continue to persist and thus pose a consid- erable challenge to both the political groupings in different countries and the EU (Mueller 2016). Many still maintain that populism is the result of dema- gogic leaders who are slightly 'evil'. However, this is not true – it is primarily a way of responding to the alienation and diffuseness, maybe even to the hyper- complexity of the globalized world in which we live. The need for identity politics, in the sense of Gemeinschaft vs Gesellschaft, provided for by populist leaders is deeply rooted in human nature (Russell Mead 2017).[11]

## Conclusion

One of the leading characteristic of modern populism is the personification of politics. In other words, parties or movements are increasingly identifiable by their founder or leader. Members and their constituency see the leader as the embodiment of the political programme and a guarantee that it will be implemented. Some analysts claim that pluralism and democracy are in danger (Mueller 2016: Mickey et al. 2017). While this may be true, in most countries with a strong populist party, the mechanisms of 'checks and balances' still exist.

The personification of politics is a necessary but not sufficient or defining factor of populism. It can be observed as a consequence of a broader process of dissatisfaction with parties and politics (in the sense of the German expression Politikverdrossenheit). It is clear that the crisis of representative (parliamentary) democracy (Mair 2013) is accelerating the shifts towards strong leadership, which may make fertile ground for populist's claims.

The other characteristic and declared intention of populism is to re-establish the sovereignty of the people and a genuine political community within national (or local) boundaries, coupled with the revolt against international connections and international institutions. However, as described in the text, this can only be done to a smaller extent, even if populists are the dominant power. This text has pointed out what the other authors and analysts of populism neglect (for instance, Mueller, 2016) – the consequences of political, cultural and economic internationalization and globalization as well as the introduction of new technological means of political communication. It can be argued that threats like the loss of national sovereignty or economic dependency can, according to the populist ideology, be solved by a strong state. A strong leader and a strong state are

---

11  It is worth quoting a wider formulation: 'Western elites believed that in the 21st century, cosmopolitanism and globalism would triumph over atavism and tribal loyalties. They failed to understand the deep roots of identity politics in the human psyche and the necessity for those roots to find political expression in both foreign and domestic policy arenas. And they failed to understand that the very forces of economic and social development that cosmopolitism and globalisation fostered would generate turbulence and eventually resistance, as Gemeinschaft (community) fought back against the onrushing Gesellschaft (market society), in the classic terms sociologists favoured a century ago' (Russell Mead 2017, p. 7).

two sides of the same coin. Put briefly: populist regimes lead inevitably to state capitalism (on the Hungarian case, see Sallai and Schnyder, 2015).

Populist politics rests on a simplified perception of societal reality. It offers simple and straight-forward recipes for how to deal with issues. As such, it does not provide productive solutions for the problems European societies are facing. However, populism cannot be perceived as the cause of these problems. It is a symptom of the crisis of established politics and the deficiencies of mainstream political elites. The latter are those who have to become able to resolve problems more effectively. Otherwise, populism of different ideological colours (or without them) will further flourish.

Let us conclude with the following thought: in order to reduce populism's attractiveness, on one the hand we should be encouraging discussions on the revitalization of democracy in both the national and EU contexts (Mueller 2016). On the other, we should be utilizing our knowledge of ways of achieving more contextual (discrete) mechanisms for regulating globalization processes. As indicated, the process of profound reformation and re-orientation at the cognitive and political levels is needed. We should take and learn the lessons from different sources, even from contemporary populism. This is the central message of this text: a phenomenon like populist politics should be studied and considered from different points of view, and various sources must be taken into account in order to avoid ideological and subjective biases. A value assessment of the meaning and consequences of populism is allowed and expected, yet it should not be the starting point of analysis.

# References

Acemoglu, Daron / Robinson A. James: Why Nations Fail? The Origins of Power, Prosperity and Poverty. London: Profile Books LTD 2013.

Author David /Dorn, David / Hanson Gordon: The China Shock: Learning from Labour Market Adjustment to large Changes in Trade. NBER Working Paper, no. 21906, MIT, Cambridge 2016.

Badiou, Alain: "Bilo je v grozi globoke noči [It was in the horror of deep night]." Mladina alternative: Posebna številka – prispevki k razumevanju časa, December 9, 2016–2017.

Beemersh, Koen /Damijan, Jože P. / Konings Jozef: Labour Market Polarization in Advanced Countries: Impact of Global Value Chains, Technology Import Competition from China and Labour Market Institutions, OECD Social, Employment and Migration Working Papers 197, OECD Publishing 2017.

Bell, Daniel A.: The China Model: Political Meritocracy and the Limits of Democracy. Princeton: Princeton University Press 2015.

Blondel, Jean/ Thiebault, Jean: Political Leadership, Parties and Citizens. The Personalization of Leadership. Oxon & New York: Routledge 2010.

Cabada, Ladislav / Tomšič, Matevž: "The Rise of person-based politics in the new democracies: the Czech Republic and Slovenia". Politics in Central Europe 12 (2), 2016, p. 29–50.

Campus, Donatella: "Mediatization and Personalization of Poltics in Italy and France: The Cases of Berlusconi and Sarkozy". International Journal of Press/ Politics 15 (2), 2010, p. 219–235.

Damijan, P. Jože: Globalizacijski šok za delovna mesta na Zahodu (Globalisation shock for jobs in the West). Svet capital (Supplement of newspaper Delo), 30. junij, 2017, p. 27.

Diamond, Larry / Richerd Gunther: "Introduction". In: Diamond, Larry / Gunther, Richard (eds.): Political parties and Democracy. Baltimore: The Johns Hopkins University Press 2001, p. ix-xxxii.

Godina, Vesna: Zablode postsocializma [Delusion of Post-socialism]. Ljubljana: Beletrina 2014.

Hopkin, Jonathan and Paolucci, Caterina: "The business firm party model of party organisation: Cases from Spain and Italy". European Journal of Political Research 35 (3), 1999, p. 307–339.

Heywood, Andrew: Political Idoelogies. An Introduction. New York: Palgrave Macmillan 2012.

Inglehart, Ronald: Cultural Shift in Advanced Industrial Society. Princeton: Princeton University Press 1990.

Irwin, Douglas: "The False Promise of Protectionism". Foreign Affairs 96 (3), 2017, p. 45–56.

Katz, Richard S. / Mair, Peter: "Changing Models of Party Organization and Party Democracy: the Emergence of the Cartel Party". Party Politics 1 (1), 1995, p. 5–31.

Keman, Hans: "Responsible Responsiveness of Parties in and out of Government". In: Harfst, Philipp / Kubbe, Ina / Poguntke,Thomas (eds.): Parties, Governments and Elites. Wiesbaden: Springer 2017, p. 25–52.

Kitschelt, Herbert: "Divergent Paths of Postcommunist democracies". In: Diamond, Larry / Gunther, Richard (eds.): Political parties and Democracy, Baltimore, The Johns Hopkins University Press 2001, p. 299–323.

Kriesi, Hans-Peter: "Personalization of national election campaigns". Party Politics 18 (6), 2011, p. 825–844.

Krouwel, André: "Party Models". In Katz, Richard S. / Crotty, William J. (eds.): Handbook of Party Politics, London: Thousand Oaks, New Delhi: Sage 2006, p. 249–269.

Laclau, Ernesto: On Populist Reason. London: Verso 2005.

Lipset, Seymour M. / Rokkan, Stein: Party Systems and Voter Alignment. New York: Free Press 1967.

Mair, Peter: Ruling the Void. New York: Verso 2013.

Makarovič, Matej / Tomšič, Matevž: "Democrats, authoritarians and nostalgics: Slovenian attitudes toward democracy". Innovative issues and approaches in social sciences 8 (3), 2015, p. 8–30.

Mancini, Paolo: Between Commodification and Lifestyle Politics: Does Silvio Berlusconi provide a new Model of Politics for Twenty-First Century? Oxford: Reuters Institute for the Study of Journalism 2011.

Mastnak, Tomaž: "Demokracija brez levice [Democracy without Left]." Mladina alternative: Posebna številka – prispevki k razumevanju časa, December 9, 2016–2017.

Mazzoleni, Oscar / Voerman, Gerrit: "Memberless parties: Beyond the business-form party model? ". Party Politics 23 (6), p. 783–792.

Meguid, Bonnie M.: Party Competition between Unequals. Strategies and Electoral Fortunes in Western Europe. Cambridge: Cambridge University Press 2007.

Menzel, Ulrich / Senghaas, Dieter: Europas Entwicklung und die Dritte Welt: Eine Bestandsaufnahme. Frankfurt: Suhrkamp 1986.

Mickey, Robert/ Levitsky, Stevem / Way, Lucan Ahmad: "Is Amerika still Safe for Democracy". Foreign Affairs 96 (3), 2017, p. 20–29.

Mudde, Cas: "The Populist Zeitgeist". Government and Opposition. 39 (4). 2004, p. 541–563.

Mudde, Cas / Kaltwasser, Cristobal R.: Populism. A Very Short Introduction. New York: Oxford University Press 2017.

Mueller, Jan W.: What is Populism? Philadelphia: University of Pennsylvania Press 2016.

Newell, James L.: The Politics of Italy. Cambridge: Cambridge University Press 2010.

Passarelli, Gianluca (ed.): The Presidentialization of Political Parties: Organizations, Institutions and Leaders. London: Palgrave Macmillan 2015.

Paulson, John: "Trump and Economy – How to Jump-Start Growth". Foreign Affairs 96 (2), 2017, p. 8–11.

Poguntke, Thomas / Webb, Paul: "The Presidentialization of Politics in Democratic Societies: A Framework for Analysis". In: Poguntke, Thomas / Webb, Paul (eds.): The Presidentialization of Politics: A Comparative Study of Modern Democracies, Oxford and New York, Oxford University Press 2005, p. 1–25.

Rizman, Rudi: "Demokracija niso le volitve in referendum [Democracy is not only election and referenda]". Mladina alternative: Posebna številka – prispevki k razumevanju časa, December 9, 2016–2017.

Rodrik, Dani: Populism and the Economics of Globalisation. Cambridge: John F. Kennedy School of Government, Harvard University 2017.

Rončević, Borut / Makarovič, Matej: "Societal steering in theoretical perspective: social becoming as an analytical solution". Polish Sociological Review 176(4), 2011, 461–472.

Rončević, Borut / Makarovič, Matej: "Towards the strategies of modern societies: systems and social processes". Innovation: The European Journal of Social Science Research, 23(3), 2010, 223–239.

Russell Mead, Walter: "The Jacksonian Revolt – American Populism and the Liberal Order". Foreign Affairs 96 (2), 2017, p. 2–7.

Sallai, Dorottya / Schnyder, Gerhard: Strong State, Weak Managers. "How Hungarian Firms Cope With Autocracy". Center for Business Research, Cambridge, Working Paper, No. 474, 2015.

Sartori, Giovani: Parties and party systems. Cambridge: Cambridge University Press 1976.

Sedláček, Tomáš: Economics of Good and Evil: The Guest for Economic Meaning from Gilgamesh to Wall Street. Oxford: Oxford University Press 2012.

Shirk, Susan: "Trump and China". Foreign Affairs 96 (2), 2017, p. 20–27.

Tomšič, Matevž: "Decline of elite consensus and destabilisation of political space in East-Central Europe". Corvinus 8 (3), 2017a, p. 151–170.

Tomšič, Matevž. "Building a common European identity: between unity and diversity". In: NovotnyÝ, Vít (ed.): Unity in adversity: immigration, minorities and religion in Europe. Brussels: Wilfried Martens Centre for European Studies 2017b, p. 29–36.

Tomšič, Matevž / Prijon Lea: "Person-based Politics in Italy and Slovenia: Comparing Cases of Leadership's Individualisation". International Social Science Journal 64 (213/214), 2013, p. 237–248.

Van Biezen, Ingrid / Poguntke, Thomas: "The decline of membership-based politics". Party Politics 20 (2), 2014, p. 205–216.

Van Zoonen Lisbet / Holtz-Bacha: "Personalisation in Dutch and German Politics: The Case of Talk Show". Javnost – The Public 7 (2), 2000, p. 45–56.

Wagner, Markus (2012): Defining and measuring niche parties, available at http://homepage.univie.ac.at/markus.wagner/Paper_nicheparties.pdf

Webb Paul /Poguntke Thomas / Kolodny Robin: "The Presidentialization of Party Leadership? Evaluating Party Leadership and Party Government in the Democratic World". In: Helms, Ludger. (ed.): Comparative Political Leadership. London: Palgrave Macmillan, 2012, p. 77–98.

Zupančič Žerdin, Alenka: "AJMO" (Let's go!). Mladina alternative: Posebna številka – prispevki k razumevanju časa, December 9, 2016–2017.

## Interviews

Laclau Ernesto, 2014, Ernesto Laclau's last Interview with La Nacion, 28. April, by D. Schinkman.

Mouffe, Chantal, 2016, America in populist Times, An Interview with Chantal Mouffe, The Nation, December, 16, by W. Shahid.

Mueller, Jan W., 2016b, Der Spuck geht nicht so schnell vorbei, in Die Zeit, nr. 5, August, 16, by K. Zeug and, N. Boeing.

Sedlaček, Tomaš, 2017, Namesto ZDA, je največji zagovornik proste trgovine Kitajska (Instead of the USA, China is now the strongest advocate of Free Trade), Delo, May 5, by N. Golc.

Žižek, Slavoj, 2016, on Arab Television Al Jazeera (http://www.youtube.com/watch?v=qhAU-6Tvo).

Dadiana Chiran

# Three Decades of Electoral Participation Between Fragmentation and Contingency: A Survey of the Literature on Economic Voting in Eastern Europe

**Abstract:** Reaching the benchmark of three decades of democracy and over nine rounds of elections represents the end of the democratic adolescence and the beginning of maturity for post-communist democracies. By summarizing eighty-five literature entries on economic voting, this review chapter focuses on a study of voting patterns and the role the economy plays in it. A first observation is that while the Western European development of the literature on economic voting has been organic, in Eastern Europe the study of economic voting was initially part of an agenda, emphasizing whether new Eastern democracies were metabolizing democracy as expected. One cause for the limited study of economic voting is a result of the lack of systematic, comparable and time-extended data, as well as the fragmentation of the post-communist region. Whether or not economic voting is the most important factor in political decision-making remains an on-going debate, although positive evidence has been identified consistently throughout the last thirty years, particularly in Hungary, Poland and, to lesser extents, in the Czech Republic and Slovakia.

**Keywords:** economic voting, CEE post-communist democracies, fragmentation, agenda-setting

## Introduction

Perhaps the most prominent way in which democratic accountability has been studied in Eastern Europe is via economic voting. Economic voting has been intensely applied in the field of behavioural and electoral studies, both in longitudinal and cross-sectional studies focusing on the electoral configurations of advanced democracies in which the data favour systematic long-term analysis. Already at the beginning of the 1990s, Western democracies were twenty-five years ahead of Central and Eastern European countries in the study of vote-popularity functions, establishing one of the most intricate dialectical discourses related to the factors that influence voting patterns. A review of the literature establishes the validity of the economic vote in advanced Western democracies as well as its robustness (for a review see Nannestad and Paldam 1993). The students of economic voting showed that variations among nations exist (see

Lewis-Beck 1986, Lewis-Beck and Stegmaier 2000, Lewis-Beck and Stegmaier 2007, Duch and Stevenson 2006), either in terms of market openness and clarity of responsibility (Powell and Whitten 1993, Hellwig and Samuels 2007, Scheve 2000, Anderson 2000), heterogeneity and level of information of voters (Duch and Palmer 2002), structural characteristics of the party system (Strom 1990) or governing coalitions (Dorussen and Taylor 2002). The extensive discourse on Western European patterns of voting testifies to an organic-driven analysis, aiming to find answers to questions such as who votes for what and why, what determines how people vote and why elections go one way or another. In answering these questions, the well-developed Western academic community appealed to a variety of methods of electoral research, from data-driven to theory-driven: direct observation of real-life (Gosnell 1930), interviews with experts (Merriam and Goswell 1924), aggregate data analysis (Siegfried 1913; Bean 1948; Budge and Farlie 1983), in-depth interviews and focus groups (Lane 1962), laboratory or real-world experiments (Gosnell 1927, Iyengar 1991) or formal modelling (Downs 1957), allowing for the formation of three primary schools of thought. The Columbia school is associated with the 'sociological model': Paul Lazarsfeld and Bernard Berelson state that the vote choice is mostly a function of social group membership. The Michigan school, associated with the '(socio)psychological model' and the American voter, elaborated by Campbell et al. (1960) indicates that vote choice is mostly an expression of long-standing identification. The Rochester school is often associated with the 'economic model', which affirms that that vote choice is a function of the spatial distance between the voter's policy preferences and candidate positions. The rational theory is an offspring of economic philosophy and was applied in political analysis by the work of Downs (1957), among others. It assumes that political decisions are made by rational minds – an extension of the homo economics concept. The logic of the transfer of the rational theory from economic to political analysis can be understood within in the context of class division as the achievement of conscious goals connected to the left-right cleavage and competitiveness of class interest. The development and application of the rational choice in Western democracies was organic, yielding push-pull effects of the national economy on voting preferences.

The fall of the Iron Curtain bestowed upon Eastern European citizens the same level of responsibility: to select and empower political leaders and consequently hold them accountable for their mandate. In this sense, the study of economic voting has become relevant in the post-communist democracies as well. This development aroused the interest of scholars in explaining the dynamics of electoral success in post-communist democracies and observing the most significant factors that influence it. Particularly because of the impact of transition on the individual economic status and overall macro-economic instability, the economic voting received special attention from the research community. The standard operationalization of economic voting has been tested in Eastern European new democracies, predominantly in country case studies (Bell 1997, Coffey 2013, Comsa and Tufis, 2014, Domonkos and Domonkos 2011, Duch 2001, Jula and Jula 2016, Przeworski 1996, Stegmaier and Lewis-Beck 2009) but also in multi-country settings (Anderson, Lewis-Beck and Stegmeier 2003, Duch 1995, Fish 1998, Fidmurc 2000, Lafay 1981, Hao Ju 2016, Harper 2000, Roberts 2008, Tucker 2006), although the level of sophistication and complexity of theoretical and empirical approaches has not (yet) come close to the Western counterpart. Although there have been voices claiming that party competition in Eastern Europe may be only marginal in structuring political attitudes (Evans and Whitefield 1993), economic voting as a basis for party competition in post-communist countries has been taken for granted by many, although the results have been somewhat incoherent, especially at early stages.

The research still lacks an extended common theoretical framework applicable to the region and a systematic analysis of the factors that explain the support for parties as a function of the economy. However, there is still time for debate, which will lead to a consensual judgment regarding the nature of economic voting in Eastern Europe; it took the Western scholarly community more than three decades of the organic development of the literature to resume a set of general conclusions regarding the parameters of economic voting: i) there is indeed a consistent economic factor influencing party success, although not the only one; ii) the VP-function stability in complex party systems is hard to maintain; iii) the cost of governance for a normal period is on average 1.7 % of the votes; iv) voters tend to be myopic (Nannestad and Paldam 1993).

This chapter focuses on the eleven Eastern European EU Member States and six candidates and how the theoretical framework of economic voting has been applied to these seventeen post-communist Central and Eastern countries and with what results or shortfalls. The chapter primarily identifies to what extent the literature suggests that the economy affects elections in the East, the main debates on the topic, as well as the consensus reached by the academic community. The chapter also presents the waves of analysis on Eastern European economic voting and whether the study of it was organic, as in the case of advanced Western democracies, or agenda-driven. The present review scrutinizes the result-oriented part of three decades of research and whether consensual conclusions regarding the intensity of economic voting can be drawn from the body of the literature on seventeen Central and East European post-communist countries (included in Tab. 1).

## The size and selection of literature

This chapter examines a body of eighty-five journal articles and book chapters, specifically (although not exhaustively) focusing on economic voting and its operationalization in Eastern Europe[1] covering the period between 1981[2] and 2018.[3] The two articles that focus on the period before the age of multiparty systems in Eastern Europe (Lafay (1981; White 1990) do not entail that the variation (or lack thereof) in communist elections could be predicted based on economic indicators, but rather that even under a political monopoly the national economy (its performance or under-performance) did influence the ministerial organigram. Their insertion into the literature sample is purely formal and serves the purpose of strengthening the explanation of fragmentation.

The primary selection criteria for the selection of literature is the analysis of economic factors in relation to election results: over 60 % of the literature analysed herewith is represented by empirical studies, whereas forty out of eighty-five articles operationalize economic voting using a variety of models, mostly based on the assumption of linearity. Table 1 presents a summary of the literature characteristics divided by the frequency of country analysis. A series of qualitative studies are also included, together with two previous reviews of voting behaviour literature (Tucker 2002 and Tucker 2015). The remaining share of the literature uses economic variables or refers to the relationship between the economy and voting patterns as a proxy for other arguments, remaining nevertheless relevant for the present endeavour. The body of the literature under review integrates studies that address electoral and vote analysis, more specifically literature in which the

---

1   The review covers seventeen Central and Eastern European countries: Albania, Bulgaria, Croatia, the Czech Republic, Estonia, Hungary, Latvia, Lithuania, Macedonia, Montenegro, Moldova, Poland, Romania, Slovakia, and Slovenia, Serbia and Ukraine. CIS, Belarus, Russia countries and the Balkan republics are only covered marginally due to the lack of literature for this region, whereas some other countries appear in comparative studies (for instance Nicaragua, Canada, Mexico or Taiwan).

2   The two articles dating from before 1990 focus on how the economic outcomes affected the succession of ministers during the communist period, more precisely whether the economy had any influence of the dismissal/nomination of ministers in the cabinet and in which countries.

3   1981 represent the date of publishing of the earliest article comprised in the review, whereas 2018 represents the latest.

electoral outputs (vote shares, turnout, vote change) and vote intentions (party preference, party success, Vote Popularity (VP) functions) are not indifferent to the fluctuations of the national economy. The common research problem of the literature herewith reviewed is how the population perceives national economy and the role that it plays in explaining both voters' choice and the party success during elections, being therefore based on three preconditions i) that information is available, ii) the vote is not constrained (democratic elections) and iii) that the underlying foundation of the electoral decision-making is indeed rationality, as assumed by the Rochester model.

To address the above-mentioned topics, the literature sample includes articles that employ quantitative analysis (specific to the positivist political theory of economic voting) and articles that, in operationalizing other concepts, also include the variables that are connected to economic voting (inflation rate, price index, unemployment rate, GDP growth rate, etc.). Both single and cross-national studies are considered in order to differentiate between the type of conclusions that they yield, whereas individual and aggregated approaches are considered based on the same rationale of observing differences and commonalities.

## The contribution of the indigenous academic community in the early stages of analysis: fragmentation explained

According to Joshua Tucker (2002: 279), there is no valid reason to expect one country receiving more attention than another and probably the most frequently analysed are either the ones that aroused the most interest to Western countries, for instance, Russia and Visegrad countries, or the ones that seemed more comparable. Similar conclusions may be drawn from the present sample of literature, although with a distinction in kind. The literature is indeed highly populated with analysis mainly focused on Central Europe, followed by the Baltic countries, Eastern countries and the Balkans are the least researched area. As indicated in Tab. 1, Hungary, Poland and the Czech Republic are by far the lead countries, both in single-country studies as well as in cross-national research. To clarify: there is only a handful of examples in which a study is conducted without these three countries partaking of it. And yet, the explanation for this regional variation cannot be accounted for only by the 'importance to the West' argument or the level of comparability. It is to be argued that the field is a positivist one; hence, it relies heavily on the availability of empirical data. A complementary explanation and one that resonates with the Western organic pattern is that the most studied countries are those with the most articulated community of scholars, research infrastructure and logistics, and consequently with an early start in electoral research: these include Hungary, Poland, Czechoslovakia and Lithuania (Evans and Whitefield 2000, Markowski 2000, Toka 2000). In Hungary, commonly known as 'the happiest barrack of Eastern Europe', electoral research began before 1990, with a few notable attempts to study the emergence of competition in the last Hungarian parliamentary elections during communism. Post-1990 Hungary was a fertile ground for survey research, both commercial and academic (see Note 6); therefore, both data and a strong academic community enhanced the study of elections, although the lack of know-how was one of the shortfalls of the process (Toka 2000:111). Especially in Hungary, the early start also yielded noteworthy comparisons, with Nicaragua, for instance, regarding the salience of ongoing politics and economics of the nation which shape the vote choice (see Anderson, Lewis-Beck and Stegmeier 2003) or comparative analysis of Hungary, Canada, the US, the Netherlands and West Germany on the subjective perception of the economy and its impact on vote choices (Duch and Palmer 2002), both based on independent surveys conducted as early as 1990.

Poland is another case in point to sustain the argument of the contribution of the indigenous scholarly community to the fragmentation of research. In Poland, political science enjoyed a fair amount of freedom in the pre-democratic era. Markowski reports that the new academic structures that were created after 1989 were dominated by sociologists, in particular, many of them prominent figures in academic life and opposition politics', delivering both in quality and quantity and 'allowing in-depth insights to be added to an already impressive body of literature' (2000:105). This insight regarding Poland and Hungary is visible in the synthetic summary of the literature (Tab. 1).

In comparison, the Baltic countries are also well represented in the literature given their size (22 %). In Lithuania's case, it is difficult to say who or what institution initiated the survey research because, in 1988, two major institutes of the University of Vilnius and Academy of Sciences existed. The first survey was conducted in 1989, and the University of Vilnius became the leader in survey research with the launch of the Lithuanian Barometer (Krupavicius 2000). Eastern countries (19 % representation within the sample) do not reach the same level of analysis. For instance, in Ukraine, Stegniy traces the first data set to 1996, highlighting that 'the empirical material is often confidential' (2000: 295). In Romania, the first traceable independent survey is through the PHARE Programme under later contract with the Soros Foundation in 1996 (Campeanu 2000). In Bulgaria, Todorov reports an early start in electoral research although 'there is no tradition of making these surveys available to academics for research purposes' (2000: 209).

**Tab. 1:** Summary of literature

| REGION | Country | No of articles: single country | Inclusion in comparative analysis (2–5 countries) | Inclusion in comparative analysis (6–10 countries) | Inclusion in comparative analysis (11 or more) | Frequency of analysis |
|---|---|---|---|---|---|---|
| **Central Europe** | - | | | | | 42% |
| 1. | Czech Rep. | 2 | 16 | 10 | 17 | 42.6% |
| 2. | Hungary | 2 | 21 | 12 | 19 | 51.3% |
| 3. | Poland | 6 | 18 | 11 | 19 | 51.3% |
| 4. | Slovakia | 2 | 15 | 11 | 16 | 41.6% |
| 5. | Slovenia | 1 | 2 | 8 | 15 | 23.9% |
| **Eastern Europe** | - | | | | | (19%) |
| 6. | Bulgaria | 1 | 3 | 11 | 17 | 29.6% |
| 7. | Romania | 3 | 3 | 13 | 16 | 32.6% |
| 8. | Moldova | 0 | 0 | 0 | 7 | 5.95% |
| 9. | Ukraine | 1 | 0 | 1 | 9 | 9.8% |
| **Baltic countries** | - | | | | | (22%) |
| 10. | Estonia | 0 | 1 | 7 | 17 | 22.6% |
| 11. | Latvia | 0 | 2 | 5 | 15 | 19.9% |
| 12. | Lithuania | 1 | 3 | 7 | 16 | 24.75% |
| **Balkan countries** | - | | | | | (6%) |
| 13. | Albania | 1 | 0 | 2 | 8 | 9.8% |
| 14. | Bosnia | 0 | 1 | 0 | 3 | 3.55% |
| 15. | Croatia | 2 | 0 | 0 | 11 | 11.3% |
| 16. | Macedonia | 1 | 1 | 0 | 8 | 8.8% |
| 17. | Serbia | 0 | 0 | 0 | 1 | 0.9% |
| | TOTAL | 27 | 26 | 13 | 19 | 85 |

* Russia, Belarus and East Germany and Yugoslavia (as a federation) are not included in the count although they are analyzed in some of the articles. Whenever Czechoslovakia is analysed in a study, both countries are counted as part of the study. Other countries that appear in the pool of studies are CIS countries, OECD countries, other Western EU countries, in some instances Taiwan, Nicaragua, the US, Canada, etc.

The Balkan peninsula represents a separate cluster mainly because in the studies on the electoral and voting patterns the emphasis is (still) often on what type of electoral system will best fit the different types of divided societies, suggesting that the literature remains stuck in structural issues and has not yet reached the subsequent phase in which, under the assumption of a stable electoral framework, scholars analyse the how and the why people form their vote choices. Even the more recent studies on voting patterns focus primarily on security issues, without including economic factors (Pickering 2009, Oliveraomar and Zivkovic 2016, Glaurdić and Vukovic 2016). Given the varieties of social stratification, ethnic structure and national identities, the Balkans are still looking for the best fit in terms of formulas for party and electoral institutionalization and consolidation. In the early 1990s, perhaps also due to social unrest, there was scarce evidence in the field of electoral studies in Croatia and only at the end of the decade did it started to produce viable datasets. Ivan Grdesic reports that 'for the lack of scholars in the field, the data will not be put to more extensive analysis and interpretation' (2000:174). In Albania, the absence of academic structures for political and sociological research is reflected in the scarcity of available studies. In 1992, the Faculty of Philosophy and Sociology was closed down, leaving a gap in research. Barjaba (2000) notes that by 2000 no research on elections or voting had been conducted in Albania, and there were no publications devoted to the subject.

To summarize, most of what we know today about economic voting in Eastern Europe is based on the evidence available from a handful of countries, either of Polish or Hungarian origin and to lesser extents Czech or Slovak. The fragmentation was shaped, among other factors, by the pace of within-country research and the earliness to gain a home-driven competitive advantage in the field. As bricks cannot be made without clay (in this case independent and available data content), the most successful countries seem to be those endowed with well-articulated academic communities, which were quick to initiate the process of research. Ultimately, we cannot fail to observe a detail that might be categorized as coincidence, namely that the countries where the oldest university centres in the East reside (specifically in Prague, Kraków, Pécs, Budapest, Bratislava and Vilnius) were those prone to kick-starting the process.

## The agenda-setting development of the study of economic voting in CEE: a different path

The organic development of Western economic voting emerged as (one of the) potential explanation for why people vote, how they arrive at their decision and what the economy can tell us about how governance will change during the next elections. According to this formal deductive theory of the Rochester model, the standard answers can be voiced as follows: vote choice is a function of the spatial distance between the voter's policy preferences and candidate positions meaning that people will vote prospectively, on an interest base, to maximize their gains. In empirical terms, economic voting deviates from the spatial voting by targeting primarily economic policy and output and is divided between the prospective as well as retrospective evaluation, under the assumption that voters arrive at their decision by being rational: the better economic performance, the more inclined voters are to reward or, otherwise, punish the leadership. In 1960, Campbell et al. found that American partisanship can be associated with economic views, whereas Fiorina (1981) concluded that voters cast a ballot in a retrospective manner and that economic judgment is sociotropic (Kiewet 1983).

The arguable difference when applying the same logic to the Eastern countries is that (operating under the same formal constraints) the study of economic voting in the East was driven by the need to assess a more urgent issue, specifically democratic accountability. Whereas the literature in the East closely followed the Western methodology in operationalizing economic voting models, it was often used to reject a different hypothesis: that the economic hardships imposed by the transition are not likely to shake the democratic convictions of the voters and that the newly emerged Eastern European democracies will not slide back into undemocratic regimes. Golob and Makarovic (2017) conclude that 'the autonomy of the political subsystem is guaranteed through a combination of a radical transition and democratic consolidation' (p.1521) although the former is not a necessary condition, suggesting why democratic consolidation received so much interest also via the analysis of economic voting in Eastern post-communist democracies. The autonomy of the economic and the political subsystems are intricate conditions for growth in human development, which further suggests the functional reasons of the fragmentation of Central and Eastern European post-communist democracies both in terms of successful transition as well as in terms of consolidation of the democratic ethos.

Separating the literature by period allows us to distinguish a few characteristics that define the development of the discourse. The literature on pre-democratic Eastern Europe represents a 'ground zero' and is limited and, therefore, underrepresented in the sample. Two scholars (Lafay 1981 and White 1990) take on the difficult task of assessing whether communist governance and intra-party dynamics are affected by economic variations, although evidence of others exists (Kukorelli 1988 in Toka 2000). Including these articles in the debate help us to understand fragmentation better and brings us closer to the agenda-setting

**Tab. 2:** Distribution of article across waves of analysis

|  | First wave (1981–1990) | Second wave (1991–1999) | Third wave (2000–2018) |
|---|---|---|---|
| No of articles that analyse 1 country | 0 | 16 | 11 |
| No of articles that analyse 2 to 5 countries | 0 | 4 | 22 |
| No of articles that analyse 6 to 10 countries | 2 | 2 | 10 |
| No of articles that analyse more than 11 countries | 0 | 2 | 16 |
| TOTAL | 2 | 24 | 59 |

argument: firstly, because this phase of pre-democratization and initial fragmentation is often omitted; secondly, since it further explains the regional variation by mapping evidence of intra-party competitiveness dictated by the variations of the economy. For pre-democratic Hungary and Poland, Jean-Dominique Lafay brings forth evidence that in those nations 'political changes respond very significantly to the internal economic situation and, to a lesser degree, to the external trade position' (1981:15). The same rule is valid to a lesser extent in Czechoslovakia, where governmental changes were severely influenced by the 1968 events; therefore, only a weak link between political changes and the internal economy can be observed (White 1990).

The distinct feature of the next wave of analysis (between 1991 and 1999) is the predominance of single-country studies, as a result of the fragmentation that characterized the literature on elections: in Hungary, for instance, the process of cooperation and scholarly exchange was initiated from 1994 onwards (Toka 2000). Poland's well-structured network of academics and supply of data is translated in extensive work (6 out of 24 the single country analyses) mostly empirical work on economic voting; noteworthy here are the important works of Bell (1997), Powers and Cox (1997), Preworski (1996) and Jackson et al. (2003[4]).

The third wave of analysis (2000–2018) is beyond doubt the most fruitful both in terms of diversity of data and methods as well as the quality and quantity of work focused on Eastern Europe. After the year 2000, the fruits of the early-formed

---

4   Jackson et al. 2003 is included in the second wave of analysis in spite of the official date of publishing would place it in the third. The incongruence lies in the detail that the article written by Jackson, Klich and Poznanska was produced in 1996–1998 and it appears in many instances as published in The British Journal of Political Science in January 1998. However the DOI indicates 2003 as official year of publishing (doi:10.1017/S0007123403000048).

science diplomacy began to deliver a considerable number of multi-country independent macro-survey projects, which were made widely available (noteworthy here are the Eurobarometer, New Democracies Barometer, Comparative Manifesto Project data, CCES data). These repositories had already hit the academic market and became part of the collective knowledge of wider communities of scholars; therefore, allowing for variation in studies to emerge, although apparently not yet in satisfactory quantities since in numerous occasions the scarcity of the literature on voting behaviours and particularly economic voting is recognized as a shortfall of the field.

Recent studies show that the party systems in the former communist countries of the EU have crystallized to the extent of being little different from those in more established democracies, in terms of providing voters with relevant electoral cues. In this sense, Der Brug, Franklin and Toka (2008) find evidence that voters and party systems in Central European countries are not (or are no longer) very different from those of Western Europe, which is not to say that party systems and political culture do not differ between old and new democracies along a number of dimensions. The same developmental approach of economic voting was theorized by Kitschelt (1992) and empirically tested by Duch (2001), Rowny and Edwards (2012) concluding that a closure of the gap between East and West is foreseen. Homogeneity within the European Union democracies is an incremental trend, although it cannot yet be predicted whether the process will be completed. Populist tendencies related to voting behaviour have also aroused as a common trait. In contemporary democracies, including Eastern Europe, the rise of populism as a response to specific political and economic crises and with the aim of defending national sovereignty (Frane and Tomsic 2019) has taken its toll in terms of economic voting by shifting the voting focus from issue competition/vote retrospection 'toward a (semi)-authoritarian (but still competitive) regime and state-led capitalism' (Frane and Tomsic 2019).

The 'agenda-setting' argument is perhaps the most visible in the literature of the nearly 1990s when democracy and whether or not it would survive under economic hardships was a joint motto among the students of economic voting. Taking a neutral stance, Herbert Kischelt (1992) hypothesizes that the absolute level of economic development determines the distribution of people along the lines of pro-market and pro-democracy. Therefore, economics ought to be seen as a predictor of democratic support. Given the hardships imposed in the early transition period, the economics of Eastern Europe could have only been a predictor of a return to authoritarian regimes. Following this line of thought, a number of researchers have argued against this grim prediction via the study of economic voting. Dutch (1995) stated that even if economic difficulties had demonstrably reduced support for governments in Russia, Hungary, Poland, the Czech Republic and Slovakia, support for democratization and institutional reform is not affected concluding that 'democracy even in nascent democracies is quite robust' (Duch 1995:152), although all of the countries in question have subsequently seen a return to power of socialist governments. Evans and Whitefield (1995) infer that:

> [...] market experience in post-communist societies, has no direct effect on democratic commitment and only has an indirect effect on democratic norms via market norms because democracy and the market have been packaged together as an ideology of opposition to communism (1995:507).

Powers and Cox (1997), using survey data from Polish elections, find that assessments on party type (former communist vs reformist) are more important than those on economic performance, suggesting that attribution of blame is directly proportional with satisfaction with reforms, regardless of economic (non-)performance. Based on data for six countries, Mishler and Rose (1996) find that despite the economic crisis 'support for the regime is based on expectations and trajectories for hope' (1996: 575), hence claiming that no matter the steepness of a declining economy, the reservoir of support for democracy is partially driven by fear of a communist regime and expectations of a hopefully better future. Comsa and Tufis's (2014) study on Romania showed that the economic crisis negatively affected confidence in political parties, confidence in political institutions and satisfaction with the functioning of democracy, but failed to have a significant effect on diffuse support for democracy.

Very often, the idea that Eastern Europeans only place their support on a dichotomous communist vs reformist scale of measurement due to lack of experience and naïve hope surfaced in the literature of the 1990s. At a subsequent stage of analysis, in a result most convincingly argued by Tucker (2006) (but also Fidrmuc, 2000, Bell 1997, Jackson et al. 2003), representing perhaps the most accurate attempt to create a theoretical frame to fit the Eastern context (discussed in more detail below), he drafted the theory of contingent economic voting, in which voters look mainly at whether a party is connected with the new or old regime rather than its responsibility for current economic performance, but without neglecting it.

The agenda-setting development, however, did not monopolize the literature entirely. Some researchers tested the relation between economic indicators and voting, among which Bell (1997), Pacek (1994), Fidmurc (2000), Harper

(2000) and Preworski (1996) with mixed results. Preworski (1996) and Bell (1997) based on data from Poland, find a strong relationship between unemployment and voting; Pacek (1994) argues that, contrary to the standard behaviour, in the case of economic failure, voters tend to abstain from voting than punish the incumbents, increasing unemployment and lowering wage levels being incentives for their withdrawal from the political process. Jan Fidmurc finds strong evidence of economic voting in all Visegrad countries; where the economic variables are seen as quite significant, the vote gain/loss of the incumbent parties being 2.8 %, twice as high as the value calculated by Paldam (1991) for Western democracies. The study of Harper (2000) stands out by the lack of significant connections in his results on Hungary, Bulgaria and Lithuania: unemployment remains statistically insignificant, concluding that:

> yet reflecting on the findings from the empirical tests of five theoretically based economic voting hypotheses, the dominant impression is that despite strong expectations to the contrary, economic factors had at best a modest effect on party preference in these societies (2000: 1225).

In comparison to Western democracies, the similarities are limited to the use of the standard methodologies. The observation that can be extrapolated from Fig. 1 is that most of the empirical studies assume a linear relationship between the response (most typically either vote share or vote choice, but also volatility, measures of trust or turnout) and the explanatory economic or additional non-economic variables. The linearity assumption was generally adopted from Western literature in which the rational choice models were tested using econometrics. The opinions on whether Western assumptions fit Eastern patters are divided: some scholars (Anderson, Lewis-Beck and Stegmeier 2003) reject the idea, whereas others are confident that convergence will occur (Kitschelt 1992).

Although the beginnings of research in the field were to some extent copy-cat works, meaning limited in terms of countries, methodology used and theoretical fit (Dassonneville and Lewis-Beck 2014, Lewis-Beck 1988), we can observe that the number of single-country and cross-national articles increase considerably with the second wave of analysis (1991–1999), in which the comparative and multi-country analyses take precedence. In the third wave of analysis (2000–2018), the modelling of the concept goes beyond the single use of OLS to more sophisticated methods of analysis, including some that break the linearity pattern, suggesting the progress made by the academic community, both in sophistication and in results.

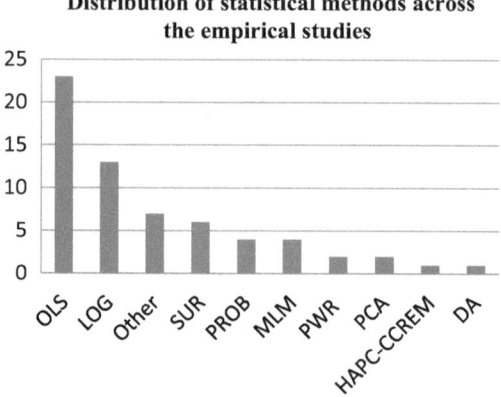

**Distribution of statistical methods across the empirical studies**

**Fig. 1:** Distribution of methods used among empirical articles. OLS – Least-squares, LOG – logistic regression; SUR – Seemingly Unrelated Regression; PROB – Probit models; MLM –Multilevel models; PWR – Prais-Winsten Regression; PCA – Principal component analysis; HAPC-CCREM – Hierarchical age-period-cohort model with cross-classified random effects; DA – Discriminant analysis

## The study of economic voting in Eastern Europe: results, debates and consensus

There is a considerable volume of theorizing and empirical evidence to suggest that voters respond to economic stimuli when they are deciding for which party they will vote. However, it is far less evident that Eastern European voters respond mechanically to economic stimuli – particularly when they arise in different political, social and cultural contexts. Political symbols, socio-demographic fabric, ethnic composition, agendas, discourses and personalities are continuously in flux. In these circumstances, it would be astonishing if identical economic stimuli produced similar reactions among voters. That is certainly not valid in the West; therefore, for all the more reason, it cannot be considered valid in the East. There is a high degree of fragmentation in the field of research on economic voting in Eastern European democracies; therefore, most of the inferences drawn are applicable only to a handful of countries. By shifting focus from results to approach, we can observe that other salient differences emerge from the comparison of Western and Eastern academic motivations of voting behaviour: whereas in the West the approach seems to be in some sense visionary (the interest is to predict (from here the rationale of using OLS) winners of future elections or simply analyse the factors that influence voters and adjust the party pledges) the study of voting behaviour in post-communist countries became more of a barometer for democracy and accountability.

Two significant lines of inquiry are identifiable in the streams of research on economic voting in Central and Eastern European democracies:

i. Classic: researchers ask whether Eastern European voters are influenced by economic factors and test the hypothesis of economic voting in Eastern Europe;

ii. Agenda-driven: even if such a pattern of voting behaviour exists, it is a mark of democratic accountability, only in correlation (not causation) with the economic factors, being therefore not expressly the result of good/bad governing.

In looking at the problem in East-West comparison, the research has been driven towards answering the puzzle of whether economic voting in the East is according to the standard model: does it present the same specificity as its Western counterpart in terms of right-left, non-incumbent vs incumbent, the salience of unemployment and similar levels of cost of governing and clarity of responsibility?

As the previous review of Tucker (2002:292) puts it 'in a way, all of [these] articles are looking for evidence to refute a null hypothesis that claims post-communist societies are so chaotic and filled with uncertainty that there will be no coherent patterns in elections and voting'. However, evidence of consensus crystallized around the macro-themes that dominated the scholarly research on economic voting in Eastern Europe.

A fragile consensus is that the economy is indeed one of the many factors that matters, although not in the same way as the standard economic model predicts. Economic voting in Eastern Europe seems to be in a permanent state of contingency, and its parameters differ from those present in the West. Several studies found that economic performance affected evaluations of incumbents: Przeworski (1996) showed a co-variation between government popularity and the unemployment rate in Poland; Duch (1995) found that economic difficulties undermine support for governments in the USSR, Hungary, Poland and the Czech Republic; Duch (2001) and Anderson, Lewis-Beck and Stegmaier (2003) found evidence of economic voting in Hungary. Using sub-national data from early in the transition period, Stegmaier and Lewis-Beck (2009) found that Hungarians acts as economic voters, following an incumbency-oriented strategy. Yet another school of thought has emerged as well. According to Pop-Eleches and Tucker (2011), 'voters look mainly at whether a party is connected with the new or old regime rather than its responsibility for current economic performance'.

Powers and Cox (1997), using cross-sectional data from the 1993 Polish election, similarly found that attitudes toward economic reforms have a limited effect on voting behaviour, but their importance is eclipsed by understandings of the past.

Some conclusions may be drawn on the impact of economic voting in the Eastern part of Europe. For the first, it was the work of Tucker (2006) that provides the most unambiguous interpretations of how economic voting explains the result of elections. Tucker's (2006) findings are consistent with Fidrmuc (2000), Jackson (et al. 2003), Mishler and Rose (1996), Powers and Cox (1997) and Anderson, Lewis-Beck and Stegmeier (2003) that the hypothesis based on party type (more accurately put, 'old regime' vs 'new regime') has more support than those based on incumbency and direct impact of the economy.

Secondly, economic voting is indeed present in Eastern Europe. The most consistent and accurate results are in Hungary (Dutch 1995, Dutch 2001, Gomez and Wilson 2006, Stegmaier and Lewis-Beck, 2009, Dutch and Palmer 2002) and Poland (Pacek 1994; Preworski 1996, Bell 1997); in recent years, other studies extended the hypothesis to multi-country dimension and found support for it (Fumarola 2016, Hao 2016, Jung 2017, Roberts 2008, Quaranta and Martini 2017). Similarly, Roberts (2008), in his cross-national study on ten Central and Eastern states, predicts that economic voting will strengthen with the passage of time. 'Citizens in new democracies can quickly learn to hold governments accountable', although their degree of accountability has to do with economic threat, not performance, as in the standard model, an argument also sustained by the work of Coffey (2013) in the Czech Republic. Roberts (2008) calculates the cost of governing

> [...] for all CEE governments, the average vote loss is 14.8 percentage points (SE 2.2). Governments in the region tend to lose five to seven times more votes than in established democracies. The previous results transfer almost in fact to the largest parties, indicating that voters hold those parties most accountable for poor economic performance (Roberts 2008: 538).

The countries in which economic voting does not seem to be applicable (or not yet revealed) are Croatia, in which the legacy of war seems to be a more efficient influencer of the pattern of voting with little variation over time (Glaurdic and Vikovic 2016) and Slovenia where voters seem to show affinity with parties reveals higher levels of ideological voting (Rovny and Edwards 2012, Jou 2011)

Another line of consensus is that unemployment matters the most (Coffee 2013, Domonkos and Domonkos 2011, Fumarola 2016, Hao 2016, Quaranta and Martini 2017, Roberts 2008, Stegmaier and Lewis-Beck, 2009, Tucker 2006), more than other economic variables such as inflation, income or productivity.

Unemployment is most sensitive for new Eastern European democracies mainly because this indicator was highly unstable after 1990 and because it was one of the communist economic dogmas most favoured by the population. Therefore, the logic is that while inflation and growth are indicators that are felt with a delay at the individual level, unemployment is a direct indicator of a minimum level of wellbeing. In the literature on economic voting, the effects of unemployment are mixed and, in general, they depend on the type of data used and the form that the dependent variable takes. With perceived survey data, unemployment comes out more strongly although there are exceptions: even though unemployment comes out as having the expected negative effect on incumbent vote shares, the results do not reach the minimum statistical significance (Harper 2000). Harper's (2000) findings regarding the failure of unemployment to reach statistical significance are at odds with Pacek's (1994) study showing that actual (i.e., non-perceived) district-level unemployment rates are related to party vote shares, but the difference in conclusions can be explained precisely by the different timeline and region, data and methodology.

## Economic voting in waves of analysis and concluding thoughts

Economic voting is, by definition, a competition-based pattern involving, at a minimum, a two-party system. However, the development of competition-based patterns can be observed in the literature prior to 1990, constituting **the first wave of analysis** (although limited). Representing only a small fraction of the literature, evidence of intra-party competitiveness and links between communist governance and economy are recorded by scholars. Lafay (1981:99) observes, based on a small dataset of UN estimates for Romania, Bulgaria, Czechoslovakia, Hungary, East Germany and Poland, that 'a politico-economic mechanism appears to be capable of explaining empirically both the economic and political fluctuations experienced in the East European countries during communism, with the addition of dummy variables for shock events' (especially in Hungary and Poland and less in Czechoslovakia, Romania and Bulgaria). The hypothesis tested is that the 'objective' conditions leading to governmental changes are essentially the following: movements in real wages, movements in consumer prices and external trade position (exports/imports); therefore, governmental changes even in one-party systems are sensitive to economic growth. In the same vein, White identifies five stages of elections in Eastern Europe: a) semi civil war; b) plebiscitary Stalinist 1920s-1950s; c) the de-stalinizing stage of the 1950s to 1960s; d) a slow widening of electoral choice between the 1960s to 1980s;

and finally e) the democratization and free election phase that started post-1990 (White: 277–280). He highlights that:

> [...] in Hungary the practice of multiple candidacies, first introduced in 1967, was made mandatory in 1983 for elections at the national level from 1985 onwards. [...] In the national elections that took place in June 1985, 42 of the 352 contestable seats failed to return a single candidate with more than 50 per cent of the vote, and 25 of the 71 independently nominated candidates were elected in place of candidates sponsored by the People's Patriotic Front (White 1990: 279).

Similar proto-intra-party competitiveness was also present in Slovenia starting the 1980s when the liberal wing of the League of Communists of Slovenia triumphed over the conservatives in the party leadership in 1986 and, consequently, the party gradually progressed towards a positive attitude in terms of political pluralism (Rizman 2006). The importance of these aspects reside in moderating the image according to which (some) Eastern voters, politicians or parties were complete novices to the notion of competitiveness, be it electoral or of other nature.

The standard interpretation is that in Eastern Europe the introduction of electoral competitiveness was under 'democracy from scratch' (Fish 1995 in Rose and Munro 2009:5) and the general view is that:

> [...] whereas previously voting had been no more than an act of mass mobilization demanded by a totalitarian (or post-totalitarian) regime to modify the ruling party, citizens are (now) faced with choices suddenly, votes represented choices between different people, parties, and movements' (Tucker 2002: 272).

This interpretation does not project an entirely accurate image of the state of facts in Eastern Europe. Similarly, it is not entirely accurate is that 'in post-Communist countries party formation commenced without civil society institutions' (Rose and Munro 2009: 22) since even during the Communist period a proto-civil society existed, although with great variations in size and activity[5] and some emerged into political formations after 1990. The role of introducing this first

---

5  For instance, Poland and the Solidarnosc labour syndicate, the Charter 77 in Czechoslovakia, the Samizdat in the entire Eastern block, the existence of Radio Free Europe and the militating of the diasporas against the communist regimes. Whereas I am not trying to suggest that civil society existed in pre-democratic Eastern Europe or that the population were politically or electorally sophisticated, the actions of silent or underground protesting against the regime as well as the limited competitiveness introduced in some national contexts are often not considered as a factor in the post 1990 voting behaviour literature, presenting Eastern Europe as *tabulae rasae*.

wave of analysis into the present review is, on the one hand, to open the appetite for further exploring government changes as a function of the economy in the East once pluralism favours the endeavour and, on the other, to contribute to the argument of research fragmentation: the split between Central European countries that better accommodate the economic model and others, such as Romania, Bulgaria and Czechoslovakia, that are less compliant (Lafay 1981: 15).

The initial applications of economic voting models between 1990 and 1999 are part of **the second wave of analysis** but represent the prima facie in the analysis of economic voting in a conditional and agenda-setting light. The effects of transition and the fear to return to autocracy were hard to ignore. Partisan attachment was difficult to achieve because of the 'missing middle': the interconnecting structures, similar to those in the West (social and political identity, participation, institutions, attachments, extensive media) which undermined the stability of voter behaviour and hindered the normative commitment to democracy (Evans and Whitefield 1993). Yet, economic voting was not hard to assume in the context. In the midst of the transition, establishing a market economy and stable institutions was achieved at a high cost. The economic burden of the market reshuffle, unformed partisanship linkages and high volatility, all compelled this wave of analysis to study less the objective dynamics between vote choice and economic outcomes and concentrate more on whether the voters' rationality is defeated by economic hardship, thus weakening the commitment to democratic values. The second wave of research concerned itself with whether: i) voters are quick to waiver their commitment to democracy in exchange of economic security and ii) are ready to exchange democracy for welfare and pro-reforms parties for old regime parties (Evans and Whitefield 1993, Pacek 1994, Duch 1995, Evans and Whitefield 1995, Mishler and Rose 1996, Powers and Cox 1997).

In his seminal work on democracy, Dahl states that democratization is 'a slow process measured in generations' (Dahl 1971: 47). In Eastern Europe, it was achieved at a rapid tempo: economic and institutional transition for most of the region was considered completed by the 2000s. At the beginning of the decade, Herbert Kischelt puts forward a number of testable hypothesis in the context of Eastern Europe (more precisely Czechoslovakia, Hungary and Poland) arguing that the absolute level of economic development determines the distribution of voters along the lines of pro-and-con market and democracy; hence, the economy becomes a proxy measurement of how much people endear democracy and the liberal market (Kitschelt 1992, Evans and Whitefield 1993). The thread of research of the decade focused on democratic accountability via economic voting but the debate whether the Western parameters would be met in the East remained. Some scholars expected Eastern Europe to converge to Western patterns on how and why people vote, how parties institutionalize and how electoral politics generally functions (Lane and Ersson 2007). As economic development continued to make incremental progress in Eastern Europe, scholars found evidence of convergence by the time of the 5th or 6th election (Van Biezen and Caramani 2007:16) or after 11 years of democratic transition (Tavits 2005).

Either as correlation or causation of this fast-forward democratization, the second wave of analysis of economic voting concentrated on weather democratization and accountability can survive in the hostile environment of transition and whether it occurs despite economic instability. In several circumstances, though, the conclusions of the articles pertaining to this wave of analysis are extrapolated beyond their empirical limits. For instance, the study of Duch (1995) based on a survey of public opinion conducted by the author with support from the USSR Institute of Sociology and an independent study conducted by Times Mirror Center in Hungary, Poland and Czechoslovakia, concludes that despite the transitional economic chaos of 1990–1991, only the incumbents suffered consequences, whereas the commitment to democratic reforms remains unshaken. The extrapolation, in this sense, intervenes in the details. Firstly, the two data sets show that, in fact, in Eastern European countries' support for democracy 'is slightly influenced by economic chaos' and 'there is some evidence that support for free markets is sensitive to economic evaluations' (Duch 1995: 152) although the model fit seems to be weaker in the Visegrad dataset than in the USSR data collection.

Moreover, in the period of 1990–1991, which is covered by the study, the 'economic chaos' had not yet reached its zenith. Yet the conclusion is somewhat relativized and slightly farfetched 'if citizens have embraced democracy, economic catastrophe is not likely to shake them of their convictions, but it will lead them to reject incumbents'. Further examples of extrapolation of conclusions, especially with regard to the validity of the statement that popular commitment to democracy is not shaken by the economic downfall of the transition, are visible in the literature.

A fully understandable explanation can be the scarcity of empirical data across the region, thus forcing the academic community to base their arguments on what was available at the time. Here, a solid connection can be made with the fact that the first set of observations regarding the existence or not of economic voting in Eastern Europe are based to a great extent on Poland and Hungary.[6]

---

6   Numerous independent surveys were commissioned in Poland and Hungary in the
    early 1990s either by state entities or private researchers under national or international
    grant financing. Considerably more than in any other Eastern European countries

Partially overlapping with Tucker's (2002) review of the literature on the broader topic of voting and elections in post-communist countries, the present chapter finds support for its initial claim that the field is indeed dominated by analyses of Poland, Hungary and the Czech Republic (Tucker 2002: 278) whereas only a few cover Eastern Europe, the Balkans or CIS states, leading therefore to the risk that most of the knowledge extracted from few contexts and the extrapolation of micro-regional conclusions to the entire post-communist area. This is, however, a trait of the second wave of analysis of economic voting, which has been corrected in the following wave of analysis that covers Eastern democracies more extensively.

The **third and latest wave of analysis** (2000–2018) is the most revealing and compact in terms of academic output and sophistication of research methods, yielding not only clearer results but also scoring high on instrumentation and revised theories that better fit the voting behaviour of Eastern Europeans. The latest wave of analysis involves a vaster and more accurate pool of methods of analysis, enlarging the number of countries analysed and including indicators which tailor the model to fit the Eastern European context: here we can mention the communist vs non-communist cleavage, the impact of transition, conflict and post-conflict political settings relevant for the Balkan states, access to information, heterogeneity, party system configurations and similar. 'Conditional Economic Voting', a term coined by Tucker, seems to be a common factor of agreement and is the most elaborate and clarifying research work within the third wave, in which the author outlines one of the most viable Eastern European voting behaviour theories. In searching for a broader model to fit the specificity of post-communist countries, Tucker moves beyond the classical approach of

---

perhaps due to the strong academic community of the two countries. A handful ought to be mentioned: 1991 an independent study conducted by Times Mirror Center; 1993 survey commissioned by the Working Group on Electoral Studies at the Institute of Political Studies, Polish Academy of Sciences, and supported by a grant from the Polish Committee of Scientific Research (KBN); 1994 Polish General Social Survey, 1994 National survey conducted by the Central European University under the coordination of Prof. Gabor Toka; monthly publications of the polish Public Opinion Research Center (CBOS) established in 1982; 1997 Hungarian Public Opinion administered by the Social Research Informatics Center of Budapest (TARKI) and funded by US National Science Foundation Grant; 1997 survey administered jointly by the University of Houston and Ezredveg Alapıtvan; the New Democracies Barometer of 1991, 1992, 1993/1994 and 1995 available for Bulgaria, Czechoslovakia (since 1992 Czech Republic and Slovakia), Hungary, Poland, Romania, Slovenia and adding five countries starting with the second study (Croatia, Ukraine, Belarus, Serbia and Moldova).

retrospective voting (in his words the Referendum Model) and explores the Transitional Model, in which New Regime parties are more likely to be favoured by good economic conditions whereas the Old Regime Parties are advantaged by poor economic conditions. He finds that the empirical winner is the Transitional model. Thus, a breakthrough empirical finding is that the effect of the economy on incumbent parties is almost entirely dependent on their relationship with the transition. Differently from Western democracies, in post-communist countries, incumbency and the right-left ideology axis are blurry (see also Kitschelt 1992).

> [The] Incumbent Hypothesis tends to dominate the way most analysts of the politics of economic reform think about elections in transition countries. And yet, we now find that in more than half of the cases for which we have empirical evidence, it is impossible to state with 90 % confidence that this relationship does exist (Tucker 2006).

There is indeed a coherent pattern of regional economic voting that links post-communist economic conditions to political outcomes that reveals a certain level of voter sophistication. Tucker finds more empirical support for the Transnational Identity model's predictions that regional economic voting patterns will be based on the relationship of political parties with the transitioning away from communism than for the Referendum model's predictions that regional economic voting patterns will be based on the party's position in or out of the government. (Tucker 2006:277).

The present endeavour indicates that research in the field of economic voting has been incrementally explanatory throughout the three waves of analysis herewith presented, although restricted to a small number of countries of Central Europe, particularly Hungary, Poland, Slovakia and the Czech Republic. The conclusions were often projected on the entire post-communist region and extrapolated beyond the research jurisdiction of the studies. One important conclusion is that there are salient differences between economic voting in the West and East. While the original development of the concept of economic voting was organic, in Eastern Europe it was initially part of an agenda, especially in the first decade of democratic rule, showing whether or not Eastern countries are metabolizing democracy as expected. Subsequently, the study of economic voting has become more revealing in Eastern Europe but without reaching the same level of sophistication and clarity. While the pattern of economic voting in new democracies is based on party type (new vs old regime) and seems to bear a cost of governance of five to seven times higher than its counterpart, in established democracies, the Referendum model prevails, and economic voting patterns are based on incumbency and a moderate level of punishment or reward.

Contributing to this literature requires a continued joint effort, especially since a considerable amount of the work in this field has focused on single countries. Comparative analysis ought to speak more straightforwardly to the discipline as a whole by presenting evidence that is less tied to limited cases, hence reducing the fragmentation of knowledge in the field, and expand beyond Visegrad countries. Although the Eastern European post-communist region is characterized by a high degree of economic, political and social fragmentation, which might suggest that the area is not sufficiently homogeneous to be studied jointly and compared as a whole to the Western model of economic voting, the globalizing factors might speak a different story. As a result of the supra-institutionalization imposed by the EU accession or future accession as well as the current internationalized trade, this review suggests an incremental pace of convergence of both voting behaviour and the impact of the economy during elections. Fragmentation in terms of the presence of economic voting also exists in advanced democracies (for instance Italy, Spain vs France, Germany or the UK) but, similar to Eastern post-communist democracies, it is a difference of degree not of kind that does not interfere in studying economic voting. The gradual yet careful inclusion into the discourse of the Balkan region (especially Croatia, Serbia, Albania, Macedonia and perhaps with more limited results Montenegro and Macedonia), as well as Eastern countries (Romania, Moldova, Ukraine, Bulgaria) and CIS countries is a must at this stage of analysis, particularly because the focus must shift from what happened in a particular election of a particular country to the study of the general trend of the voting patterns in Eastern Europe: conducting multi-country, time-series analyses on multiple elections would represent a promising start in that direction and a continuation of the organic development of the research of economic voting and voting patterns in general.

## Bibliography

Anderson, C. J. (2000). 'Economic Voting and Political Context: A Comparative Perspective'. Electoral Studies, 19 (2000), 151–170.

Anderson, L., Lewis-Beck, M. S., and Stegmaier, M. (2003). 'Post-Socialist Democratization: A Comparative Political Economy Model of the Vote for Hungary and Nicaragua', Electoral Studies, 22, 469–484.

Barjaba, K. (2000). 'Electoral research in Albania'. In H-D Klingemann, E. Mochmann and K. Newton (eds.) Elections in Central and Eastern Europe the First Wave, pp. 248–263. Berlin: WZB Sigma.

Bean, L. H. (1948). How to Predict Elections. New York: Alfred A. Knopf.

Bell, J. (1997). 'Unemployment Matters: Voting Patterns during the Economic Transition in Poland, 1990-1995'. Europe-Asia Studies, 49(7), 1263–1291.

Budge, I. and Farlie, D. J. (1983). Explaining and Predicting Elections: Issue Effects and Party Strategies in Twenty-three Democracies. London: George Allen & Unwin.

Campbell, A., Converse, P., Miller W., and Strokes, D. (1960). The American Voter. New York: Wiley.

Campeanu, P. (2000). 'Electoral research in Romania'. In H-D Klingemann, E. Mochmann and K. Newton (eds.) Elections in Central and Eastern Europe the First Wave, pp. 215 -235. Berlin: WZB Sigma

Coffey, E. (2013). 'Pain tolerance: Economic voting in the Czech Republic'. Electoral Studies, 32, 432–437. http://dx.doi.org/10.1016/j. electstud.2013.05.011.

Comsa, M. and Tufis, C. (2014). 'Reassessing the effect of economic conditions on support for democracy: Evidence from the 2009–2013 Romanian Election Study panel'. SSRN eLibrary, <https://papers.ssrn.com/sol3/papers. cfm?abstract_id=2486083>, accessed 12 February 2019.

Dahl, R. A. (1971). Polyarchy: participation and opposition. New Haven: Yale University Press

Dassonneville, R. and Lewis-Beck, M. (2014). 'Macroeconomics, Economic Crisis and Electoral Outcomes: A National European Pool', Acta Politica, 49, 372–394.

Der Brug, W. V., Franklin, M. and Toka, G. (2008). 'One Electorate or Many? Differences in Party Preference Formation between New and Established European Democracies', Electoral Studies, 27, 589–600.

Domonkos, S. and Domonkos, T. (2011). 'Economic Voting Behavior and the Political Right-Wing (Empirical Evidence from the Slovak Republic)'. Ekonomický časopis, 59(9), 905–917.

Dorussen, H. and Taylor, M. (2002). 'Group Economic Voting. A comparison of the Netherlands and Germany'. In V. Dorussen and M. Taylor (eds.), Economic Voting, pp. 92–120, London and New York: Routledge Taylor and Francis Group.

Downs, A. (1957). An economic theory of democracy. New York: Harper.

Downs, A. (1957). 'An Economic Theory of Political Action in a Democracy', Journal of Political Economy, 65, 135–150.

Duch, R. M. (1995). 'Economic Chaos and the Fragility of Democratic Transition in Former Communist Regimes', The Journal of Politics, 57(1), 121–158.

Duch, R. M. (2001). 'A Developmental Model for Heterogeneous Economic Voting in New Democracies', American Political Science Review, 95(4), 895–910.

Duch, R. M. and Palmer, H. D. (2002). 'Heterogeneous perceptions of economic conditions in cross-national perspectives". In V. Dorussen and M. Taylor (eds.). Economic Voting, pp. 139–172. London and New York: Routledge Taylor and Francis Group.

Duch, R. and Stevenson, R. (2006). 'Assessing the Magnitude of the Economic Vote over Time and across Nations'. Electoral Studies, (25), 528–547.

Evans, G. and Whitefield, S. (1993). 'Identifying the Bases of Party Competition in Eastern Europe', British Journal of Political Science, 23(4), 521–548.

Evans, G. and Whitefield, S. (1995). 'The Politics and Economics of Democratic Commitment: Support for Democracy in Transition Societies', British Journal of Political Science, 25, 485–514.

Fidmurc, J. (2000). 'Economics of Voting in Post-Communist Countries', Electoral Studies 19, 199–217.

Fiorina, M. P. (1981). 'Retrospective Voting in American National Elections'. In Gomez, B. T., and J. M. Wilson (eds.), Causal Attribution and Economic Voting in American Congressional Elections, Political Research Quarterly 56 (3): 271–282.

Fish, M. S. (1998). 'The Determinants of Economic Reform in Post-Communist World', East European Politics and Societies, 12(1), 31–78.

Frane A., and Tomsic, M. (2019). The future of populism in a comparative European and global context. Comparative sociology, ISSN 1569–1322, 2019, 18(5/6), 687–705, doi: 10.1163/15691330-12341514.

Fumarola, A. (2016). 'Much more than Economy: Assessing electoral Accountability in the CEE Member States', Politics in Central Europe, 12(2).

Golob, T. and Makarovic, M. (2017). 'Self-Organisation and Development: A Comparative Approach to Post-Communist Transformations from the Perspective of Social Systems Theory', Europe-Asia studies, ISSN 0966-8136, 69(10), 1499–1525, doi: 10.1080/09668136.2017.1399198.

Gomez, B. and Wilson, M. (2006). 'Cognitive Heterogeneity and Economic Voting: A Comparative Analysis of Four Democratic Electorates', American Journal of Political Science, 50(1), 127–145

Glaurdić, J. and Vukovic, V. (2016). 'Voting after war: Legacy of conflict and the economy as determinants of electoral support in Croatia', Electoral Studies, doi: 10.1016/j.electstud.2016.02.012.

Gosnell, H. F. (1927). Getting Out the Vote: An Experiment in the Stimulation of Voting. Chicago, IL: The University of Chicago Press.

Gosnell, H. F. (1930). Why Europe Votes? Chicago, IL: The University of Chicago Press.

Grdesic, I. (2000). 'Electoral research in Croatia'. In H-D Klingemann, E. Mochmann and K. Newton (eds.), Elections in Central and Eastern Europe the First Wave, pp. 166–181. Berlin: WZB Sigma

Hao, J. (2016). Economics and Elections: Analysis of Economic Voting in Central and Eastern European Countries during the Post-communist Era. Dissertation submitted to the Department of Political and Social Sciences of Freie Universität Berlin in fulfillment of the requirements for English Grade PhD. https://refubium.fu-berlin.de/bitstream/handle/fub188/4642/Dissertation_Ju_Hao_2016.pdf?sequence=1, accessed 23 January 2019.

Harper, M. (2000). 'Economic Voting in Post-Communist Eastern Europe', Comparative Political Studies 33(9), 1191–1227.

Hellwig, T. and Samuels, D. (2007). 'Voting in Open Economies: The Electoral Consequences of Globalisation', Comparative Political Studies, 40, 283.

Iyengar, S. (1991). Is Anyone Responsible? How Television Frames Political Issues. Chicago, IL: University of Chicago Press.

Jackson, J. A., Klich, J. and Poznanska, K. (2003). 'Democratic Institutions and Economic Reform: The Polish Case'. British Journal of Political Science, 33, 85–108.

Jou, W. (2011). 'Left–Right Orientations and Ideological Voting in New Democracies: A Case Study of Slovenia', Europe-Asia Studies, 63(1), 27–47.

Jula, N. M. and Jula, N. (2016). 'Econometric Analysis on Vote-Popularity Function for Romania, http://cks.univnt.ro/uploads/cks_2011_articles/index.php?dir=02_economy%2F&download=cks_2011_economy_art_028.pdf, accessed January 23, 2019.

Jung, D. J. (2017). 'Irrationalizing the Rational Choice Model of Voting: The Moderating Effects of Partisanship on Turnout Decisions in Western and Postcommunist Democracies', Electoral Studies, 26–38.

Kiewiet, D. R. (1983). Macroeconomics and Micropolitics: The Electoral Effects of Economic Issues. Chicago, IL: University of Chicago Press.

Kitschelt, H. (1992). 'The Formation of Party Systems in East Central Europe', Politics and Society, 7–50.

Krupavicius, A. (2000). 'Electoral research in Lithuania'. In H-D Klingemann, E. Mochmann and K. Newton (eds.), Elections in Central and Eastern Europe the First Wave, pp. 140–166 Berlin: WZB Sigma.

Lafay, J-D. (1981). 'Empirical Analysis of Politico-Economic Interaction in East European Countries', Soviet Studies, 33(3), 386–400.

Lane, J-E and Ersson, S. (2007). 'System Instability in Europe: Persistent Differences in Volatility between West and East?', Democratization, 14(1), 92–110.

Lane, R. E. (1962). Political Ideologies: Why the American Common Man Believes What He Does. New York: The Free Press.

Lewis-Beck, M. S. (1986). 'Comparative Economic Voting: Britain, France, Germany, Italy'. American Journal of Political Science, 30 (2), 315–346.

Lewis-Beck, M. S. and Stegmaier, M. (2000). 'Economic Determinants of Electoral Outcomes'. Annual Review of Political Science, 3, 183–219.

Lewis-Beck, M. S. and Stegmaier, M. (2007). 'Economic Models of Voting'. In D. Russel J. and H. D. Klingemann, The Oxford Handbook of Political Behaviour New York: Oxford University Press.

Markowski, R. (2000). 'Electoral research in Poland. In H-D Klingemann, E. Mochmann and K. Newton(eds.), Elections in Central and Eastern Europe the First Wave, pp 105–140. Berlin: WZB Sigma.

Merriam, C. E., and Gosnell, H. F. (1924). Non-Voting: Causes and Methods of Control. Chicago, IL: University of Chicago Press.

Mishler, W. and Rose, R. (1996). 'Trajectories of Fear and Hope. Support for Democracy in Post-Communist Europe', Comparative Political Studies, 28 (4), 533–581.

Mishler, W. and Rose, R. (2006). 'What Are the Origins of Political Trust?: Testing Institutional and Cultural Theories in Post-Communist Societies', Comparative Political Studies, 34 (30).

Nannestad, P. and Paldam, M. (1993). 'The VP-Function: A Survey of the Literature on Vote and Popularity Functions after 25 Years', Public Choice, 79, 213–245.

Oliveraomar, K. and Zivkovic, S. (2016). 'Montenegro: A Democracy without Alternations', East European Politics and Societies and Cultures, 30 (4), 785–804.

Pacek, A. (1994). 'Macroeconomic Conditions and Electoral Politics in East Central Europe', American Journal of Political Science, 38 (3), 723–744.

Paldam, M. (1991).'How robust is the vote function? A study of seventeen nations over four decades", in eds. H. Norpoth, M. Lewis-Beck, J-D Lafay (eds.), Economics and Politics: The Calculus of Support, pp. 9–33. Ann Arbor: University of Michigan Press.

Pickering, P. M. (2009). 'Explaining Support for Non-nationalist Parties in Post-Conflict Societies in the Balkans', Europe-Asia Studies, 61 (4), 565–591.

Pop-Eleches, G. and Tucker, J. A. (2011). 'Communism's Shadow: Postcommunist Legacies, Values, and Behavior', Comparative Politics, 43, 379–399.

Powell, B. and Whitten, G. (1993). 'A Cross-National Analysis of Economic Voting: Taking Account of the Political Context'. American Journal of Political Science, 37 (2), 391–414.

Powers, D. and Cox, J. (1997). 'Echoes from the Past: The Relationship between Satisfaction with Economic Reforms and Voting Behavior in Poland', American Political Science Review, 91(3), 617–633.

Przeworski, A. (1996). 'Public Support for Economic Reforms in Poland', Comparative Political Studies 29, 520–543.

Quaranta, M. and Martini, S. (2017). 'Does the economy really matter for satisfaction with democracy? Longitudinal and cross-country evidence from the European Union', Electoral Studies 42, 164–174.

Roberts, A. (2008). 'Hyperaccountability: Economic Voting in Central and Eastern Europe', Electoral Studies, 27, 533–546.

Rizman, R. M. (2006). Uncertain Path: Democratic Transition and Consolidation in Slovenia, Texas: A&M University Press.

Rose, R. and Munro, N. (2009). 'Parties and elections in new European democracies. Colchester: ECPR Press.

Rovny, J. and Edwards, E. (2012). 'Struggle over Dimensionality Party Competition in Western and Eastern Europe', East European Politics and Societies, 26(1), 56–74 10.1177/0888325410387635.

Sheve, K. (2000). 'Democracy and Globalization: Candidate Selection in Open Economies'. https://pdfs.semanticscholar.org/f2fb/af0fff62bbbd043 9083f0c67de90eda1303d.pdf?_ga=2.142560178.1844417733.1596880519-553190231.1596880519, accessed on 8 August 2020.

Siegfried, A. (1913). Tableau politique de la France de l'Ouest sour la Troisième République. Paris: Colin.

Stegmaier, M. and Lewis-Beck, M. (2009). 'Learning the Economic Vote: Hungarian Forecasts, 1998-2010'. Politics & Policy, 37(4), 769–780.

Stegniy, O. (2000). 'Electoral research in Ukraine'. In H-D Klingemann, E. Mochmann and K. Newton (eds.) Elections in Central and Eastern Europe the First Wave, pp. 291–303. Berlin: WZB Sigma.

Strom, K. (1990). 'A Behavioral Theory of Competitive Political Parties', American Journal of Political Science, 34 (2), 565–598.

Tavits, M. (2005). 'The Development of Stable Party Support: Electoral Dynamics in Post-Communist Europe', American Journal of Political Science, 49 (2), 283–289.

Todorov, A. (2000). 'Electoral research in Bulgaria. In H-D Klingemann, E. Mochmann and K. Newton (eds.) Elections in Central and Eastern Europe the First Wave, pp. 203–215. Berlin: WZB Sigma.

Toka, G. (2000). 'Electoral research in Hungary'. In H-D Klingemann, E. Mochmann and K. Newton (eds.), Elections in Central and Eastern Europe the First Wave, pp. 71–105 Berlin: WZB Sigma.

Tucker, J. (2002). 'The First Decade of Post-Communist Elections and Voting: What Have We Studied, and How Have We Studied It?', Annual Review Political Science, 5, 271–304.

Tucker, J. A. (2006). Regional economic voting: Russia, Poland, Hungary, Slovakia, and the Czech Republic, 1990–1999. Cambridge: Cambridge University Press.

Tucker, J. (2015). 'Comparative Opportunities: The Evolving Study of Political Behavior in Eastern Europe', East European Politics and Societies and Cultures, 29 (2), 420– 432. 10.1177/0888325414559051.

Van Biezen, I. and Caramani, D. (2007). 'Cleavage Structuring in Western vs. Central and Eastern Europe: State Formation, Nation-Building and Economic Modernization', Paper presented at the ECPR Joint Sessions, Helsinki.

White, S. (1990). 'Democratizing Eastern Europe: The Elections of 1990', Electoral Studies, 9 (4), 277–287.

# Table of Equation

# Table of Figures

# Table of Graphs

# Table of Tables

# Index

Zeitfracht Medien GmbH
Ferdinand-Jühlke-Straße 7
99095 Erfurt, Deutschland
produktsicherheit@kolibri360.de

Druck:
CPI Druckdienstleistungen GmbH
im Auftrag der
Zeitfracht Medien GmbH
Ein Unternehmen der Zeitfracht - Gruppe
Ferdinand-Jühlke-Str. 7
99095 Erfurt